Meta-Smith Charts and Their Potential Applications

Meta-Smith Charts and Their Potential Applications
Danai Torrungrueng

ISBN: 978-3-031-00411-7   paperback
ISBN: 978-3-031-01539-7   ebook

DOI 10.1007/978-3-031-01539-7

A Publication in the Springer series
SYNTHESIS LECTURES ON ANTENNAS

Lecture #10
Series Editor: Constantine A. Balanis, *Arizona State University*
Series ISSN
Synthesis Lectures on Antennas
Print 1932-6076   Electronic 1932-6084

Synthesis Lectures on Antennas

Editor
Constantine A. Balanis, *Arizona State University*

Synthesis Lectures on Antennas will publish 50- to 100-page publications on topics that include both classic and advanced antenna configurations. Each lecture covers, for that topic, the fundamental principles in a unified manner, develops underlying concepts needed for sequential material, and progresses to the more advanced designs. State-of-the-art advances made in antennas are also included. Computer software, when appropriate and available, is included for computation, visualization and design. The authors selected to write the lectures are leading experts on the subject who have extensive background in the theory, design and measurement of antenna characteristics.

The series is designed to meet the demands of 21st century technology and its advancements on antenna analysis, design and measurements for engineers, scientists, technologists and engineering managers in the fields of wireless communication, radiation, propagation, communication, navigation, radar, RF systems, remote sensing, and radio astronomy who require a better understanding of the underlying concepts, designs, advancements and applications of antennas.

Meta-Smith Charts and Their Potential Applications

Danai Torrungrueng

Asian University, Thailand

SYNTHESIS LECTURES ON ANTENNAS #10

ABSTRACT

This book presents the developments and potential applications of Meta-Smith charts, which can be applied to practical and useful transmission line problems (e.g., *metamaterial* transmission lines and *nonreciprocal* transmission lines). These problems are *beyond* the capability of the standard Smith chart to be applied effectively. As any RF engineer is aware, a key property of the Smith chart is the insight it provides, even in very complex design processes. Like the Smith chart, Meta-Smith charts provide a useful way of visualizing transmission line phenomenon. They provide useful physical insight, and they can also assist in solving related problems effectively. This book can be used as a companion guide in studying "Microwave Engineering" for senior undergraduate students as well as for graduate students. It is also recommended for researchers in the RF community, especially those working with periodic transmission line structures and metamaterial transmission lines. Problems are also provided at the end of each chapter for readers to gain a better understanding of material presented in this book.

KEYWORDS

Smith Chart, Meta-Smith Charts, transmission line, metamaterial transmission line, nonreciprocal transmission line, periodic transmission line structure, exponentially tapered nonuniform transmission line, conjugately characteristic-impedance transmission line, bi-characteristic-impedance transmission line, negative characteristic resistance, nonnegative characteristic resistance, RF circuit

*This book is dedicated to
imagination and creativity.*

*The constant support of my family,
in particular my mom and dad,
is deeply appreciated.*

Contents

Foreword I

Metamaterial microwave circuits and related printed structures have gained significant interest in recent years. Their wave slow-down properties have led to smaller matching circuits, microwave filters, feed networks and antenna arrays. Nonreciprocal propagation is another characteristic in some of these structures. With the growing interest in using metamaterials, Smith charts that can be used for designing related RF components are badly needed.

This book is a well-written text with useful problems at the end of each chapter to serve as a guide for students and instructors in using the proposed meta-charts (appropriately meaning the next-chart or after-chart upon translating the Greek word meta). The book presents meta-charts for exponentially tapered (varying characteristic impedance as a function of position along the transmission line), nonreciprocal lossless, and nonreciprocal lossy transmission lines, among others. Several applications are given, including the standard single and double stub tuners.

This is a well-written and easy-to-follow book. With the increasing use of metamaterials in design, it can be an important companion to any microwave engineering text and to users of commercial and research computational tools.

John L. Volakis
R.&L. Chope Chair Professor
The Ohio State University

Foreword II

Dramatic enhancements in technologies for manipulating materials and their properties are spurring innovations in many engineering disciplines, including those involving microwave and electromagnetic systems. The creation of new antennas, enhanced transmit and/or receive circuit devices such as switches, filters, mixers, and amplifiers, and the integration of these systems into smaller and smaller packages are all subjects of extensive current research. Achieving operation over wider bandwidths, electronically adaptable ("software defined") performance, and platform conformal integration, all at reduced cost and manufacturing complexity, are among the potential gains enabled by material innovations. I believe that the future will see our present time as the beginning of a revolution in RF technologies and the capabilities they provide due to the new materials that are currently being investigated.

While system components will be the drivers of this future, innovations in engineering design procedures will also be necessary in order to understand and to utilize the new "tools" in the engineer's toolbox that will be available. The Smith chart is one of the most fundamental tools in RF design, and accompanies the basic transmission line structures that provide connections between RF devices. As any RF engineer is aware, a key property of the Smith chart is the insight it provides in even complex design processes. Enhancement of Smith charts into "Meta-Smith charts" to enable their continued use even with transmission lines that are nonreciprocal or contain metamaterials keeps this insight available for designs using future materials. Prof. Torrungrueng makes a valuable contribution with this volume whose benefits will be realized at present, but even more so as the metamaterial revolution continues.

Joel T. Johnson
Professor, Electrical and Computer Engineering
The Ohio State University

Preface

The Smith chart was developed in early 1937 by Phillip H. Smith. In early 1939, an article on the Smith chart was first published in the *Electronics* magazine. Radio-frequency (RF) engineers have found that the Smith chart is much more than merely a graphical tool; not only does it allow the user to gain more physical insight when visualizing transmission line (TL) phenomenon, but also assists in solving associated problems effectively. Although powerful computers are used as the dominant design tool nowadays, the Smith chart still remains widely in use. It has come to provide the basis for both modern computer and measurement instrument displays.

I became impressed with the usefulness of the application of the Smith chart whilst studying courses on antennas and frequency-selective surfaces with the late Prof. Dr. Ben A. Munk during my graduate studies at The Ohio State University [1], [2]. Prof. Munk had shown that use of the Smith chart can provide a physical understanding of many associated complex phenomena. In 2002, while preparing my lecture notes on the transmission line theory from Chapter 2 in [3] (see Problem 2.29), I asked myself the question: "Can a graphical tool, like the Smith chart, be applied for nonreciprocal lossless transmission lines?" The author found that the answer to this question was "No", and so I started developing a new graphical tool for nonreciprocal lossless transmission lines. In 2004, the first journal paper on the new graphical tool, called a generalized *ZY* Smith chart or "*T-Chart*", was published [4]. I and my team have continued developing the theory of T-Charts, including their practical applications [4-10].

In the references [4-10], the authors employed the terminology "T-Charts" for graphical tools associated with conjugately characteristic-impedance transmission lines (CCITLs) and bi-characteristic-impedance transmission lines (BCITLs). These transmission lines are discussed in detail later in this book. It should be pointed out that a nonreciprocal lossless TL in [4] is an example of CCITLs. In this book, the new terminology "*Meta-Smith charts*" is used instead of "T-Charts" to make readers, who are familiar with the Smith chart, see the originality and generality of the proposed graphical tools: the Meta-Smith charts. Note that the Greek word *meta* means *beyond*. The Meta-Smith charts can be applied to more practical and useful TL problems; i.e., CCITLs and BCITLs, *beyond* the capability of the Smith chart. Another reason for using this new terminology is that the Meta-Smith charts can be applied to analyze *metamaterial TLs*, where the Smith chart cannot be employed effectively. It will be shown in the book that the Smith chart is a special case of the Meta-Smith charts.

This book is organized into five chapters and five appendices. Chapter 1 provides the essential transmission line theory. Both uniform and nonuniform transmission lines are discussed in this chapter. The material covered in this chapter will serve as an essential grounding in understanding subsequent material in the book. Chapters 2 and 3 give the theories of CCITLs and BCITLs respectively, including practical examples. These two chapters will serve as the necessary background in developing the Meta-Smith charts for CCITLs and BCITLs in Chapter 4. Chapter 5 provides several applications of the Meta-Smith charts for solving practical problems associated with CCITLs and BCITLs. Finally, five appendices are also provided to supplement material in the chapters discussed earlier. In this book, the Meta-Smith charts are applied to analyze and design passive circuits only. However, they can also be applied for problems associated with active circuits. Due to the fact that the Meta-Smith charts depend on parameters of CCITLs and BCITLs, computerized Meta-Smith chart programs are indispensable, and they are developed to generate plots of the Meta-Smith charts.

This book can be used as a companion guide in studying "Microwave Engineering" for senior undergraduate students as well as for graduate students. It is also recommended for researchers in the RF community, especially those working with periodic TL structures and metamaterial TLs. Problems are also provided at the end of each chapter for readers to gain a better understanding of material presented in this book. For convenience in usage of Meta-Smith charts, a computerized Meta-Smith chart program, developed in Java, can be accessed via `http://www.meta-smithcharts.org/`.

I hope that the Meta-Smith charts will be used extensively by students, professors and researchers in the RF community in the near future to develop a physical understanding of many sophisticated phenomena of complex transmission lines and related problems. Useful suggestions and comments from readers are welcome, and they can be submitted by e-mail to the author via `dtg@asianust.ac.th` (cc: `torrungd@hotmail.com`).

DanaiTorrungrueng
Chon Buri, Thailand

BIBLIOGRAPHY

[1] B. A. Munk, *Frequency Selective Surfaces, Theory and Design*. New York: Wiley Interscience, 2000. xvii

[2] B. A. Munk, *Finite Arrays and FSS*. New York: Wiley Interscience, 2003. xvii

[3] D. M. Pozar, *Microwave Engineering*, 2nd ed. New Jersey: Wiley, 1998. xvii

[4] D. Torrungrueng and C. Thimaporn, "A generalized *ZY* Smith chart for solvingnon reciprocal uniform transmission line problems," *Microwave and Optical Technology Letters*, vol. 40, no. 1, pp. 57–61, Jan. 2004. DOI: 10.1002/mop.11284 xvii

[5] D. Torrungrueng, P.Y. Chou, and M. Krairiksh, "An extended ZY T-chart for conjugatelycharacteristic-impedance transmission lines with active characteristic impedances," *Microwave and Optical Technology Letters*, vol. 49, no. 8, pp. 1961–1964, Aug. 2007. DOI: 10.1002/mop.22626

[6] D. Torrungrueng, P.Y. Chou, and M. Krairiksh, "A graphical tool for analysis and designof bi-characteristic-impedance transmission lines," *Microwave and Optical Technology Letters*, vol. 49, no. 10, pp. 2368–2372, Oct. 2007. DOI: 10.1002/mop.22801

[7] D. Torrungrueng, P.Y. Chou, and M. Krairiksh, "Erratum: A graphical tool for analysis and design of bi-characteristic-impedance transmission lines," *Microwave and Optical Technology Letters*, vol. 51, no. 4, pp. 1154, Apr. 2009. DOI: 10.1002/mop.24260

[8] D. Torrungrueng and C. Thimaporn, "Applications of the ZY T-chart for nonreciprocal stub tuners," *Microwave and Optical Technology Letters*, vol. 45, no. 3, pp. 259–262, May 2005. DOI: 10.1002/mop.20789

[9] D. TorrungruengandC. Thimaporn, "Application of the T-chart for solving exponentially tapered lossless nonuniform transmission line problems," *Microwave and Optical Technology Letters*, vol. 45, no. 5, pp. 402–406, Jun. 2005. DOI: 10.1002/mop.20836

[10] D. Torrungrueng, C. Thimaporn, and N. Chamnandechakun, "An application of the T-chart for solving problems associated with terminated finite lossless periodic structures," *Microwave and Optical Technology Letters*, vol. 47, no. 6, pp. 594–597, Dec. 2005. DOI: 10.1002/mop.21239

List of Abbreviations

BCITL	Bi-Characteristic-Impedance Transmission Line
CCITL	Conjugately Characteristic-Impedance Transmission Line
ETLNUTL	Exponentially Tapered Lossless Nonuniform Transmission Line
NCR	Negative Characteristic Resistance
NNCR	Nonnegative Characteristic Resistance
NRI	Negative Refractive Index
NRLSUTL	Nonreciprocal Lossy Uniform Transmission Line
NRLUTL	Nonreciprocal Lossless Uniform Transmission Line
OC	Open-Circuited
RF	Radio-Frequency
RHS	Right-Hand Side
RLSUTL	Reciprocal Lossy Uniform Transmission Line
RLUTL	Reciprocal Lossless Uniform Transmission Line
SC	Short-Circuited
TL	Transmission Line

Acknowledgments

I would like to thank my students who contributed to the development of the Meta-Smith charts; namely: Ms. Chananya Thimaporn, Mr. Po-Yen Chou, Ms. Kanokwan Vudhivorn, Ms. Natcha Chamnandechakun, Mr. Theeraputh Mekathikom, Mr. Alongkorn Darawankul, Mr. Attakorn Wongwattanarat, and Mr. Sanchai Eardprab, including my professional colleagues: Dr. Suthasinee Lamultree, Assistant Prof. Dr. Rardchawadee Silapunt, Dr. Chatrpol Lertsirimit, Assistant Prof. Dr. Chuwong Phongcharoenpanich, and Prof. Dr. Monai Krairiksh. In addition, I would like to thank Prof. Dr. Ben A. Munk of The Ohio State University for his valuable discussions leading to the discovery of the Meta-Smith charts, including his sincere encouragement and valuable suggestions for writing this book.

Furthermore, I would like to express my sincere gratitude to all reviewers for their useful feedback to improve the quality of this book; namely: Dr. Panuwat Janpugdee, Prof. Dr. Monai Krairiksh, Prof. Dr. Joel T. Johnson, Prof. Dr. John L. Volakis, Prof. Dr. Yahya Rahmat-Samii, Associate Prof. Dr. Mitchai Chongcheawchamnan, and Associate Prof. Dr. Prayoot Akkaraekthalin. Finally, I would like to thank Mr. Sanchai Eardprab for his creative artwork, including the book format arrangement.

Danai Torrungrueng
September 2010

CHAPTER 1

Essential Transmission Line Theory

1.1 INTRODUCTION

The transmission line (TL) theory is a crucial part in the general background for microwave engineers. Basically, a transmission line consists of two or more conductors, and it is employed to deliver the power from a source to a load. Figure 1.1 illustrates a two-conductor TL of length ℓ connecting the generator, with the source voltage E_S and the source impedance Z_S, to the load impedance Z_L. Typical TLs include a two-wire line, a coaxial cable and a microstrip line.

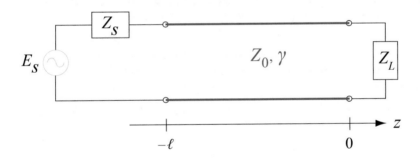

Figure 1.1: A two-conductor TL connecting the generator to the load.

A transmission line is a *distributed-parameter* two-port network, where the variations of its voltage $v(z, t)$ and current $i(z, t)$ can be observed over its length. Note that both $v(z, t)$ and $i(z, t)$ are functions of both space z and time t. TL problems can be analyzed using an extension of circuit theory or a specialization of Maxwell's equations [1]. In this chapter, the former approach is employed to study wave propagation on TLs due to its simpler mathematical formulation.

In this book, *cosine-based* phasors with $e^{j\omega t}$ time-harmonic convention are assumed in the formulation, where j is defined as $\sqrt{-1}$, $\omega = 2\pi f$ is the radian frequency, f is the frequency, and t is the time. Thus, the conversion from phasor quantities to real time-varying quantities is accomplished by multiplying the phasor by $e^{j\omega t}$ and taking the real part. For example, the real time-varying voltage $v(z, t)$ and current $i(z, t)$ can be determined from the corresponding phasor voltage $V(z)$ and phasor

current $I(z)$ as

$$v(z, t) = \text{Re}\{V(z)e^{j\omega t}\}, \tag{1.1}$$

$$i(z, t) = \text{Re}\{I(z)e^{j\omega t}\}, \tag{1.2}$$

where Re{} denotes the symbol of real part.

This chapter is organized as follows. Section 1.2 presents the TL equations derived from a lumped-element equivalent circuit model of a TL. Uniform TLs for both lossy and lossless cases terminated in a load impedance are discussed in Section 1.3. Section 1.4 presents terminated nonuniform TLs. Then, conclusions are provided in Section 1.5. Note that the material covered in this chapter will serve as an essential grounding in understanding subsequent material in the book. However, extensive treatment of the TL theory can be found in [1]–[3].

1.2   TRANSMISSION LINE EQUATIONS

Using an extension of circuit theory, a two-conductor TL of an infinitesimal length Δz can be modeled as an L-type lumped-element equivalent circuit model as shown in Figure 1.2. The TL parameters R, L, G and C are *per unit length* quantities, which are *constants* for uniform TLs. Note that R and L represent series resistance per unit length (Ω/m) and series inductance per unit length (H/m), respectively, while G and C represent shunt conductance per unit length (S/m) and shunt capacitance per unit length (F/m), respectively. Physically, R and G represent conductor loss and dielectric loss in the material between the conductors, respectively. Thus, *lossless* TLs possess $R = 0$ Ω/m and $G = 0$ S/m. It should be pointed out that other types of lumped-element equivalent circuit models can be used as well in deriving TL equations; e.g., Π-type and T-type [4].

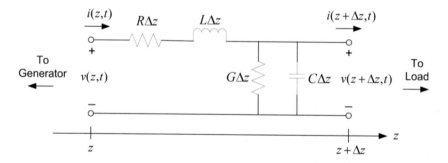

Figure 1.2: An L-type lumped-element equivalent circuit model of a two-conductor TL of an infinitesimal length Δz.

Applying Kirchhoff's voltage law and Kirchhoff's current law to the equivalent circuit model of Figure 1.2 and taking the limit as $\Delta z \to 0$ yield the following first-order coupled partial differential

TL equations in the time-domain form:

$$\frac{\partial v(z,t)}{\partial z} = -Ri(z,t) - L\frac{\partial i(z,t)}{\partial t}, \tag{1.3}$$

$$\frac{\partial i(z,t)}{\partial z} = -Gv(z,t) - C\frac{\partial v(z,t)}{\partial t}, \tag{1.4}$$

respectively.

For the sinusoidal steady-state condition, the phasor voltage $V(z)$ and phasor current $I(z)$ defined in (1.1) and (1.2) can be employed to simplify (1.3) and (1.4) considerably. Substituting (1.1) and (1.2) into (1.3) and (1.4) yields the following first-order coupled ordinary differential TL equations in the frequency-domain form:

$$\frac{dV(z)}{dz} = -(R + j\omega L)I(z), \tag{1.5}$$

$$\frac{dI(z)}{dz} = -(G + j\omega C)V(z). \tag{1.6}$$

Solving (1.5) and (1.6) simultaneously yields the following *identical* wave equations for $V(z)$ and $I(z)$:

$$\frac{d^2V(z)}{dz^2} - \gamma^2 V(z) = 0, \tag{1.7}$$

$$\frac{d^2I(z)}{dz^2} - \gamma^2 I(z) = 0, \tag{1.8}$$

where γ is the complex frequency-dependent propagation constant of the TL defined as

$$\gamma \equiv \alpha + j\beta = \sqrt{(R + j\omega L)(G + j\omega C)}. \tag{1.9}$$

Note that α and β are the attenuation constant and propagation constant of the TL, respectively. Solving (1.7) and (1.8) yields the following traveling wave solutions for $V(z)$ and $I(z)$:

$$V(z) = V_0^+ e^{-\gamma z} + V_0^- e^{\gamma z}, \tag{1.10}$$
$$I(z) = I_0^+ e^{-\gamma z} + I_0^- e^{\gamma z}. \tag{1.11}$$

In (1.10) and (1.11), the terms $e^{-\gamma z}$ and $e^{\gamma z}$ represent wave propagation in the $+z$ and $-z$ directions, respectively, for $e^{j\omega t}$ time-harmonic convention. Note that V_0^+ and V_0^- (I_0^+ and I_0^-) are defined as the amplitudes of incident and reflected voltage (current) waves referenced at $z = 0$ (at the load) respectively, (see Figure 1.1). Substituting (1.10) into (1.5), $I(z)$ can be written in terms of V_0^+ and V_0^- as

$$I(z) = \frac{1}{Z_0}\left(V_0^+ e^{-\gamma z} - V_0^- e^{\gamma z}\right), \tag{1.12}$$

where Z_0 is the complex frequency-dependent characteristic impedance of the TL defined as

$$Z_0 \equiv \frac{V_0^+}{I_0^+} = -\frac{V_0^-}{I_0^-} = \sqrt{\frac{R + j\omega L}{G + j\omega C}}. \tag{1.13}$$

It should be pointed out that the relationships between $V(z)$ and $I(z)$ are given as in (1.5) and (1.6), and in general

$$I(z) \neq \frac{V(z)}{Z_0}, \tag{1.14}$$

unless $V_0^- = 0$, (i.e., no reflected wave). Once $V(z)$ and $I(z)$ are known, $v(z, t)$ and $i(z, t)$ can be determined using (1.1) and (1.2), respectively.

For *lossless* TLs, TL parameters R and G representing loss are equal to zero. Substituting $R = G = 0$ into (1.9) and (1.13) yields

$$\alpha = 0, \tag{1.15}$$

$$\beta = \omega\sqrt{LC}, \tag{1.16}$$

$$Z_0 = \sqrt{\frac{L}{C}}. \tag{1.17}$$

Note that the attenuation constant α is zero for the lossless case as expected, and the characteristic impedance Z_0 is *purely real*. Thus, (1.10) and (1.12) can be expressed for a lossless TL as

$$V(z) = V_0^+ e^{-j\beta z} + V_0^- e^{j\beta z}, \tag{1.18}$$

$$I(z) = \frac{1}{Z_0}\left(V_0^+ e^{-j\beta z} - V_0^- e^{j\beta z}\right). \tag{1.19}$$

To determine the wavelength λ and the phase velocity v_p of waves propagating along lossless TLs, consider $v(z, t)$ in the time domain. For convenience in manipulation, V_0^+ and V_0^- can be represented in the polar form as

$$V_0^+ = |V_0^+| e^{j\phi_0^+}, \tag{1.20}$$

$$V_0^- = |V_0^-| e^{j\phi_0^-}, \tag{1.21}$$

where $\phi_0^\pm$ are the arguments of $V_0^\pm$. Substituting (1.18) into (1.1) and employing (1.20) and (1.21), $v(z, t)$ can be expressed as

$$v(z, t) = |V_0^+| \cos(\omega t - \beta z + \phi_0^+) + |V_0^-| \cos(\omega t + \beta z + \phi_0^-). \tag{1.22}$$

As pointed out earlier, the first term represents a wave propagating in the $+z$ direction, while the second term represents a wave propagating in the $-z$ direction. Without loss of generality, consider only the first term in determining the wavelength λ and the phase velocity v_p. By definition, the wavelength is defined as the distance between two successive reference point on the wave, *at a fixed instant of time*, possessing the phase difference of 2π. Using this definition, λ can be expressed in terms of β as

$$\lambda = \frac{2\pi}{\beta}. \tag{1.23}$$

For the phase velocity, it is defined as the velocity at which a *fixed phase point* on the wave travels. Consider a fixed phase point ϕ_c = constant; i.e.,

$$\omega t - \beta z = \phi_c. \tag{1.24}$$

Using (1.24) and (1.23), the phase velocity v_p can be determined as

$$v_p = \frac{dz}{dt} = \frac{\omega}{\beta} = f\lambda. \tag{1.25}$$

For lossless TLs, substituting (1.16) into (1.25) yields

$$v_p = \frac{1}{\sqrt{LC}}. \tag{1.26}$$

1.3   TERMINATED UNIFORM TRANSMISSION LINES

In this section, effects of load termination on uniform TLs will be studied for both lossy and lossless cases. Figure 1.3 illustrates a uniform TL terminated in an arbitrary load impedance Z_L. Only *passive* load impedances ($\text{Re}\{Z_L\} \geq 0$) are of interest in this section.

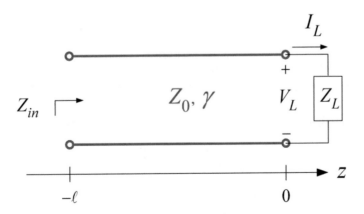

Figure 1.3: A uniform TL terminated in a load impedance Z_L.

1.3.1   TERMINATED LOSSY UNIFORM TRANSMISSION LINES

Lossy TLs were studied intensively in [5]. For lossy uniform TLs, the complex propagation constant γ and the characteristic impedance Z_0 are *complex* quantities. Let us define the voltage reflection coefficient Γ at the load as the ratio of the amplitude of the reflected voltage wave to that of incident one; i.e.,

$$\Gamma \equiv \frac{V_0^-}{V_0^+}. \tag{1.27}$$

Using (1.27), (1.10), and (1.12) can be rewritten in terms of Γ as

$$V(z) = V_0^+ \left(e^{-\gamma z} + \Gamma e^{\gamma z}\right), \tag{1.28}$$

$$I(z) = \frac{V_0^+}{Z_0} \left(e^{-\gamma z} - \Gamma e^{\gamma z}\right), \tag{1.29}$$

where Γ can be determined by considering the load condition; i.e.,

$$Z_L \equiv \frac{V(z = 0)}{I(z = 0)} = Z_0 \left(\frac{1 + \Gamma}{1 - \Gamma}\right). \tag{1.30}$$

Solving (1.30) for Γ yields

$$\Gamma = \frac{Z_L - Z_0}{Z_L + Z_0}. \tag{1.31}$$

Note that $\Gamma = 0$ when the load impedance Z_L is equal to the characteristic impedance Z_0. Under this condition, the TL *matches* to the load with *no reflection* of the incident wave as can be seen from (1.28) and (1.29) with $\Gamma = 0$. In general, the voltage reflection coefficient $\Gamma(z= -d)$ at an arbitrary location $z = -d$ along the TL can be defined as the ratio of the reflected component to the incident one, (see (1.28)); i.e.,

$$\Gamma(z = -d) \equiv \frac{V_0^+ \Gamma e^{-\gamma d}}{V_0^+ e^{\gamma d}} = \Gamma e^{-2\gamma d}. \tag{1.32}$$

Note that $\Gamma(z=-d)$ and Γ are closely related as given in (1.32), and $\Gamma(z=-d)$ is equal to zero when $\Gamma = 0$ as expected.

The input impedance Z_{in} as shown in Figure 1.3 can be determined from (1.28) and (1.29) as

$$Z_{in} \equiv \frac{V(z = -\ell)}{I(z = -\ell)} = Z_0 \left(\frac{1 + \Gamma e^{-2\gamma \ell}}{1 - \Gamma e^{-2\gamma \ell}}\right). \tag{1.33}$$

Using (1.32), (1.33) can be rewritten compactly in terms of the voltage reflection coefficient Γ_{in} at the input, where $\Gamma_{in} = \Gamma(z = -\ell)$, as

$$Z_{in} = Z_0 \left(\frac{1 + \Gamma_{in}}{1 - \Gamma_{in}}\right). \tag{1.34}$$

For convenience in calculation, Z_{in} in (1.33) can alternatively be expressed in terms of the load impedance Z_L, by using (1.31), as

$$Z_{in} = Z_0 \left(\frac{Z_L + Z_0 \tanh{(\gamma \ell)}}{Z_0 + Z_L \tanh{(\gamma \ell)}} \right). \tag{1.35}$$

Comparing (1.30) and (1.34), it is obvious that they are in the identical form, which is useful for constructing a graphical tool, called *the Smith chart* [3], for solving *reciprocal uniform* TL problems (see Problem 1.3 at the end of this chapter). It should be pointed out that the Smith chart is more than a graphical tool. It usually provides a useful way with more physical insight of visualizing TL phenomenon and solving TL problems effectively [1]. Extensive discussion and useful applications of the Smith chart can be found in detail in [3].

It is interesting to point out that the magnitude of the voltage reflection coefficient at the load $|\Gamma|$ can *exceed unity* for reciprocal *lossy* uniform TLs (RLSUTLs) terminated by a *passive* load impedance Z_L. This can be seen from (1.31) as follows. Let us define the normalized impedance z as

$$z \equiv \frac{Z}{Z_0}, \tag{1.36}$$

where Z is the impedance of interest. Using (1.36), Γ in (1.31) can be rewritten compactly as:

$$\Gamma = \frac{z_L - 1}{z_L + 1}, \tag{1.37}$$

where z_L is the normalized load impedance. Since the characteristic impedance Z_0 and the load impedance Z_L are *complex* quantities in general, let us conveniently define them in the polar form as:

$$Z_0 = |Z_0| e^{-j\phi_0}, \tag{1.38}$$

$$Z_L = |Z_L| e^{j\phi_L}, \tag{1.39}$$

where $|Z_0|$ and $-\phi_0$ are the magnitude and the argument of Z_0 respectively, and $|Z_L|$ and ϕ_L represent the magnitude and the argument of Z_L, respectively. Since all parameters of TLs in (1.13) are *nonnegative*, it is obvious that the angle ϕ_0 must lie in the following range:

$$-45° \leq \phi_0 \leq 45°. \tag{1.40}$$

In addition, for a *passive* load, (1.39) implies that

$$-90° \leq \phi_L \leq 90°. \tag{1.41}$$

Let us define the normalized load impedance z_L as:

$$z_L = r_L + jx_L, \tag{1.42}$$

where r_L and x_L are the real and imaginary parts of z_L, respectively. From (1.36), it should be pointed out that the arguments of the load impedance Z_L and the normalized load impedance z_L are generally different since the characteristic impedance Z_0 is a *complex* quantity. For example, when Z_L is real and Z_0 is complex, the normalized load impedance z_L is complex. Using (1.36), (1.38), and (1.39), it can be shown readily that [6]

$$r_L = \left| \frac{Z_L}{Z_0} \right| \cos(\phi_L + \phi_0). \tag{1.43}$$

Note that if $\cos(\phi_L + \phi_0) < 0$, (e.g., $\phi_L = 90°$ and $\phi_0 = 30°$), r_L will become a *negative* quantity. Using (1.37) and (1.42), the magnitude of the voltage reflection coefficient at the load $|\Gamma|$ can be written as:

$$|\Gamma| = \sqrt{\frac{(r_L - 1)^2 + x_L^2}{(r_L + 1)^2 + x_L^2}}. \tag{1.44}$$

From (1.44), it is obvious that when $r_L < 0$, $|\Gamma|$ *can exceed unity* since the additional term x_L^2 is the same for both numerator and denominator.

1.3.2   TERMINATED LOSSLESS UNIFORM TRANSMISSION LINES

As pointed out earlier in Section 1.2, the attenuation constant α of lossless TLs is zero; i.e., the complex propagation constant γ is *purely imaginary*, and its characteristic impedance Z_0 is *purely real*. Note that a lossless TL is a special case of a lossy TL when the loss is vanished. Following the same analysis as in the lossy case, the voltage reflection coefficient Γ at the load for the lossless case is given as in (1.31). In addition, the input impedance Z_{in} for the lossless case is given as in (1.33), where γ is equal to $j\beta$. For convenience in calculation, Z_{in} in (1.35) for the lossless case can be expressed in terms of the load impedance Z_L as

$$Z_{in} = Z_0 \left(\frac{Z_L + jZ_0 \tan(\beta\ell)}{Z_0 + jZ_L \tan(\beta\ell)} \right). \tag{1.45}$$

For reciprocal *lossless* uniform TLs terminated by a *passive* load impedance Z_L, the magnitude of the voltage reflection coefficient at the load $|\Gamma|$ cannot *exceed unity* ($|\Gamma| \leq 1$). This can be seen by considering (1.44), and using the fact that r_L defined in (1.43) is always greater than or equal to zero ($r_L \geq 0$) for the lossless case ($\phi_0 = 0°$ in (1.38) and ϕ_L is given in (1.41)). In the next section, terminated nonuniform TLs are presented.

1.4   TERMINATED NONUNIFORM TRANSMISSION LINES

Nonuniform transmission lines (NUTLs) have several advantages over uniform ones, and they have found various applications in microwave technology [7]–[13]; i.e., impedance transformers, resonators, delay equalizers, directional couplers and filters. For applications on impedance transformers, NUTLs are generally used for matching two unequal impedances over a wide band of frequencies. The reflection coefficient of voltage/current along NUTLs can be described by a non-linear Ricatti-type differential equation, and its general solution does not exist analytically [13].

To illustrate analytical procedures for analyzing NUTLs, a simple and well-known NUTL is considered in the following subsection; i.e., the exponentially tapered lossless nonuniform transmission line (ETLNUTL). To be practically useful, ETLNUTLs are considered in the presence of passive load terminations.

1.4.1 TERMINATED EXPONENTIALLY TAPERED LOSSLESS NONUNIFORM TRANSMISSION LINES

One of the NUTLs, the ETLNUTL, has found various applications in microwave technology; e.g., a resonator, an antenna, and a matching device suitable for matching two unequal impedances over a wideband of frequencies [14]–[17]. By definition, an ETLNUTL is a TL whose impedance per unit length $Z_\ell(z)$ and admittance per unit length $Y_\ell(z)$ vary *exponentially* with distance down the line. Note that its impedance and admittance per unit length can be expressed mathematically as [15]

$$Z_\ell(z) = j\omega L_0 e^{qz}, \tag{1.46}$$
$$Y_\ell(z) = j\omega C_0 e^{-qz}, \tag{1.47}$$

respectively. Note that L_0 and C_0 are the inductance and capacitance per unit length at the load ($z = 0$) respectively, and q is the *real* taper factor, which can be positive, zero or negative. In Figure 1.4, the ETLNUTL of length ℓ is employed to match a *passive real* load impedance Z_L to the *lossless* feed line of characteristic impedance Z_0, where Z_0 is *real*.

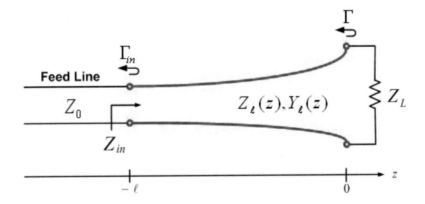

Figure 1.4: An ETLNUTL of length ℓ is employed to match a passive real load impedance Z_L to the lossless feed line of characteristic impedance Z_0.

For an ETLNUTL, the characteristic impedance $Z_c(z)$ along the ETLNUTL is defined as [15]

$$Z_c(z) = \sqrt{\frac{Z_\ell(z)}{Y_\ell(z)}}. \tag{1.48}$$

Substituting (1.46) and (1.47) into (1.48), the characteristic impedance $Z_c(z)$ can be expressed as

$$Z_c(z) = Z_c(z=0)e^{qz} = \sqrt{\frac{L_0}{C_0}}e^{qz}. \tag{1.49}$$

To match the load to the feed line, the characteristic impedances at the input ($z = -\ell$) and at the load ($z = 0$) are required as follows, (see Figure 1.4):

$$Z_c(z=-\ell) = e^{-q\ell}Z_c(z=0) = Z_0, \tag{1.50}$$
$$Z_c(z=0) = Z_L. \tag{1.51}$$

Note that the phasor voltage $V(z)$ and the phasor current $I(z)$ distributed along the ETLNUTL can be determined exactly by solving the following *first-order coupled* ordinary differential equations [14], [15]:

$$\frac{dV(z)}{dz} = -Z_\ell(z)I(z), \tag{1.52}$$

$$\frac{dI(z)}{dz} = -Y_\ell(z)V(z). \tag{1.53}$$

Note that (1.52) and (1.53) can be rearranged into the set of the *second-order uncoupled* ordinary differential equations as follows:

$$\frac{d^2V(z)}{dz^2} - \left[\frac{1}{Z_\ell(z)}\frac{dZ_\ell(z)}{dz}\right]\frac{dV(z)}{dz} - [Z_\ell(z)Y_\ell(z)]\,V(z) = 0, \tag{1.54}$$

$$\frac{d^2I(z)}{dz^2} - \left[\frac{1}{Y_\ell(z)}\frac{dY_\ell(z)}{dz}\right]\frac{dI(z)}{dz} - [Z_\ell(z)Y_\ell(z)]\,I(z) = 0. \tag{1.55}$$

Using (1.46) and (1.47), (1.54) and (1.55) can be simplified as

$$\frac{d^2V(z)}{dz^2} - q\frac{dV(z)}{dz} + \omega^2 L_0 C_0 V(z) = 0, \tag{1.56}$$

$$\frac{d^2I(z)}{dz^2} + q\frac{dI(z)}{dz} + \omega^2 L_0 C_0 I(z) = 0, \tag{1.57}$$

which are the wave equations of $V(z)$ and $I(z)$ for ETLNUTLs, respectively. Unlike the wave equations for uniform TLs, (see (1.7) and (1.8)), (1.56) and (1.57) are *different*. Solving (1.56) and (1.57) analytically, it is found that the phasor voltage $V(z)$ and the phasor current $I(z)$ can be written compactly in terms of the superposition of two traveling waves propagating in the opposite directions as follows:

$$V(z) = V_1 e^{-\gamma_{V,1} z} + V_2 e^{-\gamma_{V,2} z}, \tag{1.58}$$
$$I(z) = I_1 e^{-\gamma_{I,1} z} + I_2 e^{-\gamma_{I,2} z}, \tag{1.59}$$

where the subscripts 1 and 2 are associated with waves propagating in the $+z$ and $-z$ directions, respectively, and

$$\gamma_{V,1} = -\frac{q}{2} + j\beta, \tag{1.60}$$

$$\gamma_{I,1} = \frac{q}{2} + j\beta. \tag{1.61}$$

Note that $\gamma_{V,2}$ and $\gamma_{I,2}$ in (1.58) and (1.59) are *complex conjugate* of $\gamma_{V,1}$ and $\gamma_{I,1}$, respectively, and the propagation constant β is defined as

$$\beta \equiv \sqrt{\omega^2 L_0 C_0 - \left(\frac{q}{2}\right)^2}. \tag{1.62}$$

For convenience in analysis, β is assumed to be *real*, and only *nonnegative* values of β are considered. In (1.58) and (1.59), V_i and I_i are voltage and current amplitudes of propagating waves referenced at the load respectively, where $i = 1$ and 2. In (1.58), it is observed that the magnitudes of the first and second terms on the right-hand side (RHS) of the phasor-voltage equation increase as z increases. On the contrary, the magnitudes of the first and second terms on the RHS of the phasor-current equation, (see (1.59)) decrease as z increases. However, it can be shown rigorously that the *time-average real power* is *constant* along the ETLNUTL for *real q* and *nonnegative β*; i.e., there is *no power loss* when waves propagate along the ETLNUTL. This result agrees with the fact that the ETLNUTL is *lossless*.

Next, substituting (1.58) and (1.59) into (1.52) and (1.53), one obtains the following relationships between V_i and I_i ($i = 1$ and 2):

$$I_1 = \frac{V_1}{Z_0^+}, \tag{1.63}$$

$$I_2 = -\frac{V_2}{Z_0^-}, \tag{1.64}$$

where $Z_0^{\pm}$ are the *effective* characteristic impedances of ETLNUTLs defined as

$$Z_0^+ \equiv \frac{\gamma_{I,1}}{j\omega C_0}, \tag{1.65}$$

$$Z_0^- \equiv -\frac{\gamma_{I,2}}{j\omega C_0}. \tag{1.66}$$

For real q and nonnegative β, it can be shown rigorously that Z_0^+ and Z_0^- are related as follows:

$$Z_0^+ = \left(Z_0^-\right)^*, \tag{1.67}$$

where the superscript "*" denotes the complex conjugate symbol. Note that the total voltage $V(z = 0)$ and current $I(z = 0)$ at the load are related to the load impedance Z_L as

$$Z_L = \frac{V(z = 0)}{I(z = 0)}. \tag{1.68}$$

The voltage reflection coefficient Γ *at the load* for ETLNUTLs is defined as:

$$\Gamma \equiv \frac{V_2}{V_1}. \tag{1.69}$$

Using (1.58), (1.59), (1.63), (1.64), (1.68), and (1.69), Γ can be expressed compactly in terms of Z_0^+, Z_0^-, and Z_L as

$$\Gamma = \frac{Z_L Z_0^- - Z_0^+ Z_0^-}{Z_L Z_0^+ + Z_0^+ Z_0^-}. \tag{1.70}$$

Next, let us consider the input impedance Z_{in} of ETLNUTLs as shown in Figure 1.4, which is defined as

$$Z_{in} \equiv \frac{V(z = -\ell)}{I(z = -\ell)}. \tag{1.71}$$

Substituting (1.58) and (1.59) into (1.71) and using (1.60), (1.61), (1.63), (1.64), and (1.69), Z_{in} can be expressed compactly as

$$Z_{in} = Z_0^+ Z_0^- \left[\frac{1 + \Gamma e^{-j2\beta\ell}}{Z_0^- - Z_0^+ \Gamma e^{-j2\beta\ell}}\right] e^{-q\ell}. \tag{1.72}$$

To verify that the expression of Z_{in} in (1.72) is valid, one can compute the *input* reflection coefficient Γ_{in} from Z_{in} in (1.72), where Γ_{in} is defined as, (see Figure 1.4)

$$\Gamma_{in} \equiv \frac{Z_{in} - Z_0}{Z_{in} + Z_0}, \tag{1.73}$$

and compare the above result to the analytical solution for Γ_{in} given in [13]. To achieve this, it is more convenient to express Z_{in} in the following form:

$$Z_{in} = \left(\frac{Z_L \left(Z_0^+ + Z_0^-\right) + j\left[Z_L \left(Z_0^+ - Z_0^-\right) + 2Z_0^+ Z_0^-\right] \tan \beta\ell}{Z_0^- + Z_0^+ + j\left(Z_0^- - Z_0^+ + 2Z_L\right) \tan \beta\ell}\right) e^{-q\ell}. \tag{1.74}$$

Note that (1.74) is obtained by substituting Γ in (1.70) into (1.72) and rearranging terms. Finally, substituting (1.74) into (1.73) and manipulating terms, Γ_{in} can be written compactly as

$$\Gamma_{in} = \frac{q \sin \beta\ell}{2\beta \cos \beta\ell + j2\omega\sqrt{L_0 C_0} \sin \beta\ell}. \tag{1.75}$$

It is found that the resulting input reflection coefficient in (1.75) is identical to the existing one in [13]. Thus, it can be concluded that the derived input impedance Z_{in} in (1.72) is valid indeed.

1.5   CONCLUSIONS

In this chapter, the covered material provides a necessary grounding in understanding material in subsequent chapters, including mathematical notations and conventions employed in the book. Both uniform and nonuniform TLs are discussed in detail, and associated TL equations and parameters are obtained. The formulas of two important parameters, the voltage reflection coefficient and the input impedance, are derived for both uniform and nonuniform TLs. These parameters will be reconsidered when constructing and applying associated Meta-Smith charts in Chapters 4 and 5.

BIBLIOGRAPHY

[1] D. M. Pozar, *Microwave Engineering*, 3rd ed. New Jersey: Wiley, 2005. 1, 2, 7

[2] U. S. Inan and A. S. Inan, *Engineering Electromagnetics*. California: Addison Wesley Longman, 1999.

[3] P. H. Smith, *Electronic Applications of the Smith Chart*. Georgia: Noble Publishing, 2000. 2, 7

[4] M.N.O. Sadiku, *Elements of Electromagnetics*, 4th ed. New York: Oxford University Press, 2007. 2

[5] F. E.Gardiol, *Lossy Transmission Lines*. Massachusetts: Artech House, 1987. 6

[6] D. Torrungrueng, A. Wongwattanarat, and M. Krairiksh, "Magnitude of the voltage reflection coefficient of terminated reciprocal uniform lossy transmission lines,"*Microwave and Optical Technology Letters*, vol. 49, no. 7, pp. 1516–1519, Jul. 2007. DOI: 10.1002/mop.22490 8

[7] R. W. Klopfenstein, "A transmission line taper of improved design," *Proc. IRE*, vol. 44, pp. 31–35, Jan. 1956. DOI: 10.1109/JRPROC.1956.274847 8

[8] D. Youla, "Analysis and synthesis of arbitrarily lossless nonuniform lines," *IEEE Trans. on Circuit Theory*, vol. 12, no. 1, pp. 145–145, Mar. 1965.

[9] E. N. Protonotarios and O. Wing, "Analysis and intrinsic properties of the general nonuniform transmission line," *IEEE Trans. on Microwave Theory and Tech.*, vol. MTT-15, no. 3, pp. 142–150, Mar. 1967. DOI: 10.1109/TMTT.1967.1126403

[10] A. Bergquist,"Wave propagation on nonuniform transmission lines," *IEEE Trans. On Microwave Theory and Tech.*, vol. MTT-20, no. 8, pp. 557–558, Aug. 1972. DOI: 10.1109/TMTT.1972.1127809

[11] S. Uysal, *Nonuniform Line Microstrip Directional Couplers and Filters*. Massachusetts:Artech House, 1993.

[12] K. Lu, "An efficient method for analysis of arbitrary nonuniform transmission lines," *IEEE Trans. on Microwave Theory and Tech.*, vol. MTT-45, no. 1, pp. 9–14, Jan. 1997. DOI: 10.1109/22.552026

[13] R. E. Collin, *Foundations for Microwave Engineering*, 2nd ed. New York: McGraw-Hill, 1992. 8, 9, 12

[14] R. N.Ghose, "Exponential transmission lines as resonators and transformers," *IRE Transactions on Microwave Theory and Tech.*, vol. 5, no. 3, pp. 213–217, Jul. 1957. DOI: 10.1109/TMTT.1957.1125143 9, 10

[15] N. H. Younan, B. L. Cox, C. D. Taylor, andW.D. Prather, "An exponentially tapered transmission line antenna," *IEEE Transactions on Electromagnetic Compatibility*, vol. 36, no. 2, pp. 141–144, May 1994. DOI: 10.1109/15.293276 9, 10

[16] S. P. Mathur and A. K. Sinha, "Design of microstrip exponentially tapered lines to match helical antennas to standard coaxial transmission lines," *IEE Proceedings H on Microwaves, Antennas and Propagation*, vol.135, no. 4, pp. 272 – 274, Aug. 1988.

[17] D. Torrungrueng and C. Thimaporn, "Application of the T-chart for solving exponentially tapered lossless nonuniform transmission line problems," *Microwave and Optical Technology Letters*, vol. 45, no. 5, pp. 402–406, Jun. 2005. DOI: 10.1002/mop.20836 9

1.6   PROBLEMS

1.1. Derive the partial differential TL equations of (1.3) and (1.4) by using Kirchhoff's voltage law and Kirchhoff's current law to the equivalent circuit model of Figure 1.2 and taking the limit as $\Delta z \to 0$.

1.2. Show that the input impedance Z_{in} of a terminated lossy uniform transmission lines as shown in Figure 1.3 is given as in (1.35).

1.3. The Smith chart is a polar plot of the voltage reflection coefficient Γ at the load of terminated uniform transmission lines. It can be employed to convert from voltage reflection coefficients to *normalized* impedances (or admittances), and vice versa. Using (1.31) and (1.36), show that the equations of *resistance and reactance circles* in the Smith chart are given as

$$\left(\Gamma_r - \frac{r_L}{1+r_L}\right)^2 + \Gamma_i^2 = \left(\frac{1}{1+r_L}\right)^2,$$

$$(\Gamma_r - 1)^2 + \left(\Gamma_i - \frac{1}{x_L}\right)^2 = \left(\frac{1}{x_L}\right)^2,$$

respectively. Note that Γ_r and Γ_i are the real and imaginary parts of Γ, and r_L and x_L are the real and imaginary parts of the normalized load impedance z_L defined in (1.42), respectively. In addition, plot the families of resistance and reactance circles for several values of r_L and x_L in the Γ plane, respectively. Do the resistance and reactance circles intersect at the right angle? Please explain in detail.

Using the same concept as above, derive the equations of *conductance and susceptance circles* in the Smith chart as well. In addition, plot the families of conductance and susceptance circles for several values of g_L and b_L in the Γ plane, where g_L and b_L are the real and imaginary parts of the normalized load admittance $y_L = 1/z_L$, respectively.

1.4. Using the fact that all parameters of TLs in (1.13) are *nonnegative*, show that the angle ϕ_0 of Z_0, (see (1.38)) must lie in the range defined in (1.40).

1.5. Using (1.52) and (1.53), derive (1.54) and (1.55) rigorously. In addition, show that the phasor voltage $V(z)$ and the phasor current $I(z)$ given in (1.58) and (1.59) are the solutions of (1.56) and (1.57), respectively.

1.6. Using (1.58) and (1.59), show that the *time-average real power* $P_{av}(z)$ along the ETLNUTL, defined as

$$P_{av}(z) = \frac{1}{2}\mathrm{Re}\left\{V(z)I^*(z)\right\},$$

is *constant* for *real q* and *nonnegative* β, (see (1.60) to (1.62)). Does this result make physical sense? Please explain. In addition, derive (1.67) for *real q* and *nonnegative* β.

1.7. Show that the voltage reflection coefficient Γ at the load and the input impedance Z_{in} of terminated ETLNUTLs are given in (1.70) and (1.72), respectively. Determine the condition when terminated ETLNUTLs have no reflected wave. What is the input impedance Z_{in} under this condition?

1.8. Rigorously derive (1.74) and (1.75) for terminated ETLNUTLs.

CHAPTER 2

Theory of Conjugately Characteristic-Impedance Transmission Lines (CCITLs)

2.1 INTRODUCTION

A conjugately characteristic-impedance transmission line (CCITL) is a class of transmission lines (TLs) used in microwave technology. By definition, CCITLs are *lossless* and possess the *complex-conjugate* characteristic impedances, Z_0^+ and Z_0^-, of waves propagating in *forward* and *reverse* directions, respectively; i.e., $Z_0^+ = (Z_0^-)^*$. In general, CCITLs can possess *different* propagation constants β^+ and β^- for propagation in the forward and reverse directions, respectively, as shown in Figure 2.1; i.e., CCITLs can be *nonreciprocal* transmission lines as well. Note that β^+ and β^- are *real* while Z_0^+ and Z_0^- are generally *complex*.

In Figure 2.1, the CCITL is terminated in a passive load impedance Z_L ($\text{Re}\{Z_L\} \geq 0$), where Γ is the voltage reflection coefficient at the load and Z_{in} is the input impedance of the terminated CCITL. Examples of CCITLs are nonreciprocal lossless uniform transmission lines (NRLUTLs), reciprocal lossless uniform transmission lines (RLUTLs), exponentially tapered lossless nonuniform transmission lines (ETLNUTLs) and finite reciprocal lossless periodic TL structures operated in passbands. These TLs will be discussed in detail in Sections 2.2 to 2.5.

In practice, CCITLs can exhibit *nonnegative characteristic resistances* (NNCRs) and *negative characteristic resistances* (NCRs); i.e., $\text{Re}\{Z_0^\pm\} \geq 0$ and $\text{Re}\{Z_0^\pm\} < 0$, respectively. For example, finite reciprocal lossless periodic TL structures can exhibit both NNCRs and NCRs in passbands, which will be described in more detail in Sections 2.5, 5.4, and 5.5 [1], [2].

This chapter is organized as follows. Section 2.2 presents nonreciprocal lossless uniform transmission lines as the first example of CCITLs. Reciprocal lossless uniform transmission lines as a special case of CCITLs are discussed in Section 2.3, and exponentially tapered lossless nonuniform transmission lines are presented in Section 2.4. Section 2.5 shows the theory of finite reciprocal lossless periodic TL structures and illustrates that they can be analyzed using reciprocal CCITLs. Finally, conclusions are provided in Section 2.6.

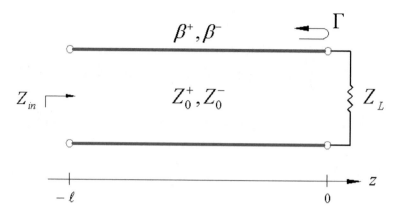

Figure 2.1: A nonreciprocal CCITL terminated in a passive load Z_L.

2.2   NONRECIPROCAL LOSSLESS UNIFORM TRANSMISSION LINES

Nonreciprocal TLs have found various applications in microwave technology; e.g., wave guiding structures, microstrip and lens antennas, and reflectionless coating. Note that a nonreciprocal TL has *different* TL parameters for waves propagating in *opposite* directions. The concept of nonreciprocal TLs is particularly useful when analyzing electromagnetic wave propagation in layered structures of bi-isotropic media [3]. The theory of nonreciprocal TLs was studied thoroughly in [3], [4].

Figure 2.1 illustrates a model of a nonreciprocal TL of length ℓ terminated in a passive load impedance Z_L. Note that the nonreciprocal TL possesses the different propagation constants β^+ and β^-, with corresponding characteristic impedances, Z_0^+ and Z_0^-, for propagation in the *forward* and *reverse* directions, respectively. In Appendix B, it is shown that $\beta^+ \neq \beta^-$ is the required condition for *nonreciprocal lossless* uniform TLs. On the nonreciprocal TL, the traveling wave equations for the phasor voltage $V(z)$ and the phasor current $I(z)$ can be written as

$$V(z) = V_0^+ e^{-j\beta^+ z} + V_0^- e^{j\beta^- z}, \tag{2.1}$$

$$I(z) = \frac{V_0^+}{Z_0^+} e^{-j\beta^+ z} - \frac{V_0^-}{Z_0^-} e^{j\beta^- z}, \tag{2.2}$$

where $e^{-j\beta^+ z}$ and $e^{j\beta^- z}$ terms represent waves propagating in the $+z$ and $-z$ directions, respectively. Note that V_0^+ and V_0^- are defined as the amplitudes of incident and reflected voltage waves referenced at $z = 0$, respectively.

For a nonreciprocal *lossless* uniform transmission line, the time-average power must be *constant* along the line. Using this condition, it can be shown rigorously in Appendix C that the characteristic impedances $Z_0^{\pm}$ must be *complex conjugate* of one another, which can be expressed as

$$Z_0^+ = \left(Z_0^-\right)^*. \tag{2.3}$$

Thus, NRLUTLs are *nonreciprocal* CCITLs. For convenience in manipulation later, let us define $Z_0^{\pm}$ in the polar form as

$$Z_0^{\pm} = \left|Z_0^{\pm}\right| e^{\mp j\phi}, \tag{2.4}$$

where $|Z_0^{\pm}|$ and ϕ are the absolute value and the argument of Z_0^-, respectively.

Note that the *total* voltage and current at the load ($V(0)$ and $I(0)$) are related by the load impedance Z_L as follows:

$$Z_L = \frac{V(0)}{I(0)}. \tag{2.5}$$

The voltage reflection coefficient Γ at the load is defined as:

$$\Gamma \equiv \frac{V_0^-}{V_0^+}. \tag{2.6}$$

Using (2.1), (2.2), (2.5), and (2.6), Γ can be mathematically expressed in terms of Z_L and Z_0 as

$$\Gamma = \frac{Z_L Z_0^- - Z_0^+ Z_0^-}{Z_L Z_0^+ + Z_0^+ Z_0^-}. \tag{2.7}$$

In Figure 2.1, the input impedance at the point $z = -\ell$ is given by

$$Z_{in} = \frac{V(z = -\ell)}{I(z = -\ell)} = Z_0^+ Z_0^- \frac{1 + \Gamma e^{-j2\tilde{\beta}\ell}}{Z_0^- - Z_0^+ \Gamma e^{-j2\tilde{\beta}\ell}}, \tag{2.8}$$

where (2.1), (2.2), and (2.6) are employed. Note that $\tilde{\beta}$ in (2.8) is defined as the *arithmetic* mean of $\beta^{\pm}$, which can be expressed as

$$\tilde{\beta} \equiv \frac{1}{2}\left(\beta^+ + \beta^-\right) \equiv \frac{2\pi}{\tilde{\lambda}}, \tag{2.9}$$

where $\tilde{\lambda}$ is defined as the *effective* wavelength of waves propagating along NRLUTLs. In the next section, reciprocal lossless uniform TLs are discussed as an example of CCITLs.

2.3   RECIPROCAL LOSSLESS UNIFORM TRANSMISSION LINES

Reciprocal lossless uniform transmission lines are discussed earlier in Section 1.3.2, where their complex propagation constant $\gamma = j\beta$ is *purely imaginary*, and their characteristic impedance Z_0 is *purely real*. The phasor voltage $V(z)$ and phasor current $I(z)$ of RLUTLs are given as, (see (1.18) and (1.19))

$$V(z) = V_0^+ e^{-j\beta z} + V_0^- e^{j\beta z}, \tag{2.10}$$

$$I(z) = \frac{1}{Z_0}\left(V_0^+ e^{-j\beta z} - V_0^- e^{j\beta z}\right), \tag{2.11}$$

where V_0^+ and V_0^- are defined as the amplitudes of incident and reflected voltage waves, respectively. Comparing (2.10) and (2.11) to (2.1) and (2.2) for NRLUTLs, it is found that the RLUTL is a special case of NRLUTLs when $Z_0^+ = Z_0^- = Z_0$ and $\beta^+ = \beta^- = \beta$. It should be pointed out that (2.7) and (2.8) (Γ and Z_{in} for NRLUTLs) are reduced to (1.31) and (1.33) (Γ and Z_{in} for RLUTLs with $\gamma = j\beta$), respectively, when $Z_0^+ = Z_0^- = Z_0$ and $\beta^+ = \beta^- = \beta$. Note that $Z_0^+ = Z_0^- = Z_0$ (which is purely real) implies that $Z_0^+ = (Z_0^-)^*$. Thus, the RLUTL is a special case of CCITLs with the identical and real Z_0^+ and Z_0^-; i.e., the CCITL is a generalization of the RLUTL, which allow us to handle broader class of useful TL problems. In the next section, exponentially tapered lossless nonuniform TLs are revisited (discussed earlier in Section 1.4.1).

2.4   EXPONENTIALLY TAPERED LOSSLESS NONUNIFORM TRANSMISSION LINES

In Section 1.4.1, it is found that the phasor voltage $V(z)$ and the phasor current $I(z)$ can be written explicitly in terms of the superposition of two traveling waves propagating in the opposite directions as follows (see (1.58) to (1.61), (1.63), and (1.64)):

$$V(z) = \left(V_1 e^{-j\beta z} + V_2 e^{j\beta z}\right)e^{\frac{qz}{2}}, \tag{2.12}$$

$$I(z) = \left(\frac{V_1}{Z_0^+}e^{-j\beta z} - \frac{V_2}{Z_0^-}e^{j\beta z}\right)e^{-\frac{qz}{2}}, \tag{2.13}$$

where the subscripts 1 and 2 are associated with waves propagating in the $+z$ and $-z$ directions, respectively, and (see (1.62), (1.65), and (1.67))

$$\beta = \sqrt{\omega^2 L_0 C_0 - \left(\frac{q}{2}\right)^2}, \tag{2.14}$$

$$Z_0^+ = \frac{\gamma_{I,1}}{j\omega C_0} = \left(Z_0^-\right)^*. \tag{2.15}$$

Note that L_0 and C_0 are the inductance and capacitance per unit length at the load, respectively, and $\gamma_{I,1}$ is defined as in (1.61). For convenience in analysis, β is assumed to be *real*, and only *nonnegative* values of β are considered. This can be achieved when the real taper factor q is chosen as $q \leq 2\omega\sqrt{L_0 C_0}$, (see (2.14)). From (2.12), (2.13), and (2.15), it is found that ETLNUTLs are *reciprocal* CCITLs with NNCRs for $q \leq 2\omega\sqrt{L_0 C_0}$. In the next section, finite reciprocal lossless periodic TL structures are discussed as a useful and interesting example of CCITLs.

2.5 FINITE RECIPROCAL LOSSLESS PERIODIC TRANSMISSION LINE STRUCTURES

Periodic structures of TLs have several applications in microwave technology; e.g., microwave filters, slow-wave components, traveling-wave amplifier, phase shifters, and antennas. A periodic structure usually refers to an *infinite* TL or waveguide periodically *loaded* with reactive elements [5]. Periodic TL structures support *slow-wave* propagation, and they possess *passband* and *stopband* character-istics similar to those of filters. However, periodic structures are always *finite* in length in practical applications. Effects of finite size of periodic TL structures and load termination will be studied in this section. It should be pointed out that periodically loaded TLs can also be designed to simul-taneously exhibit *effective negative refractive index* (NRI) and *negative group delay* in the frequency bands of interest by using a resonant circuit embedded within each loaded TL unit cell, which are *metamaterial TLs* having useful engineering applications [6], [7].

In this section, only *reciprocal lossless uniform* TLs are considered as elements of each unit cell of periodic TL structures. It is assumed that periodic TL structures are operated in *passbands* only, where nonattenuating and propagating waves exist on periodic TL structures [5]. Each unit cell of periodic TL structures can be symmetrical or unsymmetrical as well.

Figure 2.2 illustrates an example of an *unsymmetrical* unit cell of a lossless periodic TL struc-ture. The unit cell consists of a reciprocal lossless uniform TL of length d loaded with lossless lumped elements (C_1, L, C and L_1) across the midpoint of the line. It is assumed that these lumped elements have no physical length at the frequency band of interest. Note that this *lossless* unit cell is similar to that in [6] for metamaterial TLs (see Figure 1 in [6]), which simultaneously exhibits effective NRI and negative group delay. It should be pointed out that this periodic TL structure of the unit cell in Figure 2.2 is *dispersive*. The characteristic impedance and the propagation constant of the *unloaded* uniform TL are defined as Z_1 and k, respectively, as shown in Figure 2.2.

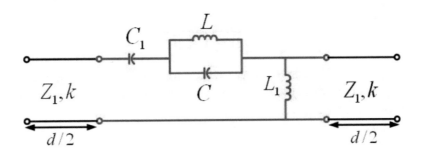

Figure 2.2: An example of an *unsymmetrical* unit cell of a lossless periodic TL structure.

The traveling wave equations for the phasor voltage V_m and the phasor current I_m *in the passbands* at the terminal of the mth unit cell of terminated reciprocal lossless periodic TL structures as shown in Figure 2.3 (where $m = 1, 2, \ldots, M$) can be written compactly as [5]

$$V_m = V_0^+ e^{-jm\beta d} + V_0^- e^{jm\beta d}, \tag{2.16}$$

$$I_m = \frac{V_0^+}{Z_B^+} e^{-jm\beta d} + \frac{V_0^-}{Z_B^-} e^{jm\beta d}, \tag{2.17}$$

where V_0^+ and V_0^- are defined as the amplitudes of the incident and reflected voltage waves referenced at the input of the periodic TL structure, respectively, and $Z_B^\pm$ are referred to as the *Bloch impedances*. Note that V_0 and I_0 in Figure 2.3 denote the total input voltage and the total input current of the

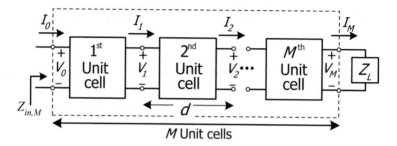

Figure 2.3: A finite reciprocal lossless periodic TL structure of M unit cells terminated in a load impedance Z_L.

terminated finite reciprocal lossless periodic TL structure of M unit cells, respectively. V_0 and I_0 can be obtained from (2.16) and (2.17) by letting $m = 0$, respectively. Using the *ABCD* matrix technique (see Appendix A) and *eigenanalysis* [5], it can be shown rigorously that the Bloch impedances $Z_B^\pm$ can be expressed in terms of the *total ABCD* parameters as

$$Z_B^\pm = \frac{-2B}{A - D \mp \sqrt{(A + D)^2 - 4}}, \tag{2.18}$$

where Z_B^+ and Z_B^- are the Bloch impedances for forward and reverse traveling waves, respectively. From Figure 2.2, the total *ABCD* matrix of the unsymmetrical unit cell can be calculated by using the cascading property of the *ABCD* matrix [5].

In the *passband* operation, it is required that [5]

$$|A + D| \leq 2. \tag{2.19}$$

Thus, it is more convenient to express $Z_B^\pm$ in (2.18) for the passband operation as

$$Z_B^\pm = \frac{-2B}{A - D \mp j\sqrt{4 - (A + D)^2}}. \tag{2.20}$$

In (2.16) and (2.17), β is the *effective* propagation constant of the periodic TL structure operating in passbands, and it can be determined from the following dispersion relation

$$\cos \beta d = \frac{A + D}{2}, \tag{2.21}$$

which is a result of eigenanalysis of periodic TL structures [5]. Using (2.19), the magnitude of the right-hand side of (2.21) is always less than or equal to 1.0 in passbands; i.e., β is always *real* in passbands.

For a *reciprocal lossless* unit cell, it can be shown rigorously that its *total ABCD* matrix possesses the following properties; i.e., A and D are *purely real* and B and C are *purely imaginary*. In addition, for a *symmetric* unit cell, it can be shown rigorously that A and D are *identical* ($A = D$). Using (2.20) and (2.21), the Bloch impedances $Z_B^{\pm}$ and the effective propagation constant β of the periodic TL structure with *symmetric* unit cells *operating in passbands* satisfy the following equations:

$$Z_B^{\pm} = \frac{\mp jB}{\sqrt{1 - A^2}}, \tag{2.22}$$

$$\cos \beta d = A, \tag{2.23}$$

where $A \leq 1$ (see (2.19)). It is observed that $Z_B^{\pm}$ are *purely real* with *different signs* for *reciprocal lossless symmetric* unit cells *operating in passbands* since B is *purely imaginary* and $A \leq 1$.

In practice, periodic TL structures are finite in size, and they are usually connected to a load impedance. Let us consider a finite reciprocal lossless periodic TL structure of M unit cells terminated in a load impedance Z_L as illustrated in Figure 2.3. The input impedance $Z_{in,M}$ at the input terminal of the terminated reciprocal lossless periodic TL structure can be expressed compactly as (see the concept in [8])

$$Z_{in,M} \equiv \frac{V_0}{I_0} = Z_B^+ Z_B^- \frac{1 + \Gamma e^{-j2M\beta d}}{Z_B^- + Z_B^+ \Gamma e^{-j2M\beta d}}, \tag{2.24}$$

where Γ is the voltage reflection coefficient *at the load*, which is mathematically given as

$$\Gamma \equiv \frac{V_0^- e^{jM\beta d}}{V_0^+ e^{-jM\beta d}} = -\frac{Z_L Z_B^- - Z_B^+ Z_B^-}{Z_L Z_B^+ - Z_B^+ Z_B^-}. \tag{2.25}$$

Next, it will be shown that *reciprocal* CCITLs can be used to analyze finite reciprocal lossless periodic TL structures.

Figure 2.4 illustrates a *reciprocal* CCITL, terminated in a passive load impedance Z_L, possessing the unique propagation constant $\beta (\beta^+ = \beta^- = \beta$ in (2.1) and (2.2)) with conjugate characteristic impedances, Z_0^+ and Z_0^-, for propagation in the *forward* and *reverse* directions, respectively. Using (2.8) and (2.9) with $\beta^+ = \beta^- = \beta$, the input impedance Z_{in} of the terminated reciprocal CCITL can be written as

$$Z_{in} = Z_0^+ Z_0^- \frac{1 + \Gamma e^{-j2\beta \ell}}{Z_0^- - Z_0^+ \Gamma e^{-j2\beta \ell}}, \tag{2.26}$$

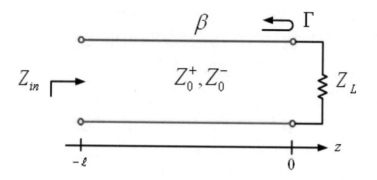

Figure 2.4: A reciprocal CCITL terminated in a passive load impedance Z_L.

where Γ is the voltage reflection coefficient *at the load* defined in (2.7). Comparing (2.24) to (2.26), $Z_B^{\pm}$ and $Z_0^{\pm}$ are related as

$$Z_B^{\pm} = \pm Z_0^{\pm}. \tag{2.27}$$

It should be pointed out that, with the aid of (2.27), the expressions of Γ defined in (2.7) and (2.25) are identical. Using (2.24) and (2.27), the input impedance $Z_{in,M}$ at the input terminal of the terminated periodic TL structure shown in Figure 2.3 can be rewritten in terms of $Z_0^{\pm}$ as

$$Z_{in,M} = Z_0^{+} Z_0^{-} \frac{1 + \Gamma e^{-j2M\beta d}}{Z_0^{-} - Z_0^{+}\Gamma e^{-j2M\beta d}}, \tag{2.28}$$

where $Z_0^{\pm}$ are defined as the *effective* characteristic impedances of the periodic TL structure. From (2.28), it can be concluded that the equation of the input impedance of terminated finite reciprocal lossless periodic TL structures is the same as that of terminated reciprocal CCITLs even $M = 1$ (one unit cell). Thus, these periodic TL structures *when considering at the input terminal of each unit cell* can be effectively analyzed using reciprocal CCITLs. Due to flexibility in fabricating periodic TL structures and their usefulness in practice, there are several applications in microwave technology as pointed out earlier. It should be pointed out that lossless periodic TL structures are probably the most useful example of CCITLs in practice.

From (2.20), (2.21), and (2.27), the effective characteristic impedances $Z_0^{\pm}$ and the effective propagation constant β of the reciprocal lossless periodic TL structure are generally dependent on various parameters of each unit cell of the reciprocal lossless periodic TL structure through the total *ABCD* parameters of each unit cell in a complicated fashion. As pointed out earlier in Section 2.2, the characteristic impedances $Z_0^{\pm}$ of CCITLs must be *complex conjugate* of one another as shown in (2.3). For convenience, $Z_0^{\pm}$ are defined in a polar form as shown in (2.4). It is interesting to point out that reciprocal lossless periodic TL structures can exhibit both NNCRs ($\text{Re}\{Z_0^{\pm}\} \geq 0$) and NCRs ($\text{Re}\{Z_0^{\pm}\} < 0$) in passbands as shown in an example below. For periodic TL structures with NNCRs, (2.4) implies that the argument ϕ of Z_0^- must lie in the following range:

$$-90° \leq \phi \leq 90°. \tag{2.29}$$

However, the argument ϕ for periodic TL structures with NCRs must lie in the following ranges:

$$-180° < \phi < -90° \cup 90° < \phi \leq 180°, \tag{2.30}$$

where the symbol "$\cup$" denotes the union in set theory. Note that, for periodic TL structures with *reciprocal lossless symmetric* unit cells *operating in passbands*, $Z_0^\pm$ are *identical* and *purely real*; i.e., the argument ϕ is equal to 0° for NNCRs or 180° for NCRs (see (2.22) and (2.27)).

From (2.28) and (2.7), it is obvious that the *matching condition* resulting in $Z_{in,M} = Z_L$ for terminated periodic TL structures can be obtained when $Z_L = Z_0^+$ for the NNCR case and $Z_L = -Z_0^-$ for the NCR case; i.e., the magnitude of the voltage reflection coefficient at the load, $|\Gamma|$, is equal to zero and approaches infinity, respectively. It should be pointed out that $Z_L = -Z_0^-$ for the NCR case can be realizable with *passive* loads since $\text{Re}\{Z_0^-\} < 0$. Appendix D shows the derivation of $|\Gamma|$ for both NNCR and NCR cases. It is interestingly found that $|\Gamma|$ is always less than or equal to unity for the NNCR case, and always greater than or equal to unity for the NCR case. However, associated powers for both cases are still conserved for *passive* load terminations as shown in detail in Appendix E.

An example of periodic TL structures exhibiting both NNCRs and NCRs is illustrated in Figure 2.5, where the number of unit cells is equal to one ($M = 1$) for simplicity. The lossless *unsymmetrical* unit cell is connected with a passive load and a generator with the source voltage E_s and the source impedance Z_s. The unit cell is implemented by using capacitively loaded reciprocal lossless uniform TLs with the propagation constant k and the characteristic impedance Z_1, where $d/2$ is the length of each uniform TL as shown in Figure 2.5. In this example, $d = 2$ cm, $Z_1 = 150\,\Omega$, $C_s = 5.6$

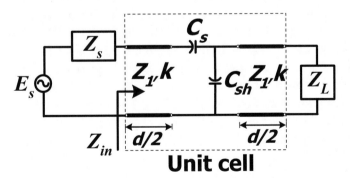

Figure 2.5: A lossless unsymmetrical unit cell connected with a passive load and a generator.

pF, and $C_{sh} = 0.44$ pF are employed. In this example, it is assumed that the phase velocity of wave propagating along the uniform TLs is equal to 3×10^8 m/s. In addition, $Z_L = 200\,\Omega$ is employed, and the frequency of interest ranges from 0.5 to 10 GHz. The passbands can be observed when $|\cos\beta d| \leq 1$ as shown in Figure 2.6. It is found that there are two passbands in this frequency range; i.e., the 1st passband (0.67–4.15 GHz) and the 2nd passband (7.56–9.73 GHz). Between the 1st and 2nd passbands, a stopband exists in the frequency range between 4.15 GHz and 7.56 GHz, where $|\cos\beta d| \geq 1$ is observed.

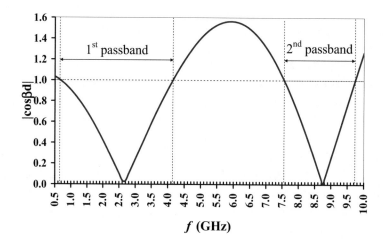

Figure 2.6: The plot of the magnitude of $\cos\beta d$ versus frequency.

Figures 2.7 and 2.8 plot the magnitudes and the arguments of $Z_0^{\pm}$ versus frequency, respectively. Note that ϕ^+ and ϕ^- in Figure 2.8 are the arguments of Z_0^+ and Z_0^- respectively, and ϕ^- is the same as ϕ defined in (2.4). From Figures 2.7 and 2.8, it is observed that Z_0^+ and Z_0^- are *complex conjugate* of each other in the passbands only as expected; i.e., $|Z_0^+| = |Z_0^-|$. From Figure 2.8, it is found that ϕ^- is in the ranges given in (2.29) and (2.30) for the 1st and 2nd passbands, respectively; i.e., the periodic TL structure ($M \geq 1$) for the unit cell given in Figure 2.5 exhibits NNCRs in the 1st passband and NCRs in the 2nd passband. Away from the passbands, (i.e., stopbands), $Z_0^{\pm}$ are *purely imaginary* with *different* magnitudes. Figure 2.9 illustrates the plot of the magnitude of Γ versus frequency. From the plot, it is found that $|\Gamma| \leq 1$ and $|\Gamma| \geq 1$, corresponding to NNCRs and NCRs, are observed in the 1st and 2nd passbands, respectively. These results agree very well with the derivation given in Appendix D.

2.6   CONCLUSIONS

In this chapter, the theory of CCITLs is discussed in detail. Practical examples of CCITLs are also provided, and lossless periodic TL structures are probably the most useful example of CCITLs. It is interestingly found that CCITLs can exhibit nonnegative characteristic resistances ($\text{Re}\{Z_0^{\pm}\} \geq 0$) and negative characteristic resistances ($\text{Re}\{Z_0^{\pm}\} < 0$). It is rigorously shown that the magnitude of

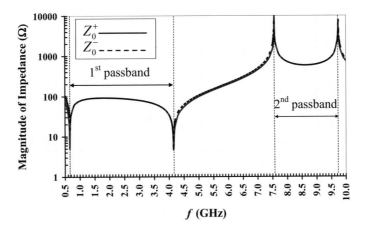

Figure 2.7: The plot of the magnitudes of Z_0^+ and Z_0^- versus frequency.

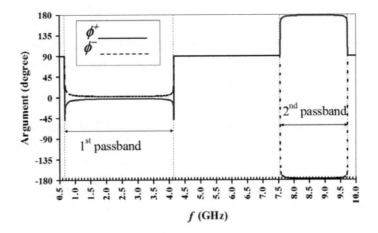

Figure 2.8: The plot of ϕ^+ and ϕ^- versus frequency.

the voltage reflection coefficient at the load is always less than or equal to unity for the NNCR case, and always greater than or equal to unity for the NCR case. However, associated powers for both cases are still conserved for passive load terminations. To gain more physical understanding and simplify the analysis and design of CCITLs, graphical tools based on the Smith chart will be developed in Chapter 4. In the next chapter, CCITLs will be generalized to include effects of loss of associated TLs and/or lumped elements.

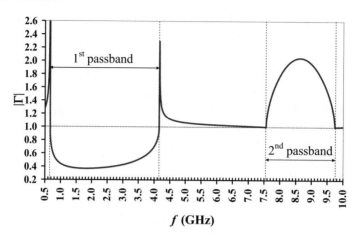

Figure 2.9: The plot of the magnitude of Γ versus frequency.

BIBLIOGRAPHY

[1] S. Lamultree and D. Torrungrueng, "On the characteristics of conjugately characteristic-impedance transmission lines with active characteristic impedance," in *2006 Asia-Pacific Microwave Conference Proceedings*, Dec. 2006, vol. 1, pp. 225–228. DOI: 10.1109/APMC.2006.4429411 17

[2] D. Torrungrueng, S. Lamultree, C. Phongcharoenpanich and M. Krairiksh, "In-depth analysis of reciprocal periodic structures of transmission lines," *IET Microwaves, Antennas & Propagation*, vol. 3, no. 4, pp. 591–600, Jun. 2009. DOI: 10.1049/iet-map.2008.0205 17

[3] I. V. Lindell, M. Valtonen, and A. H. Sihvola, "Theory of nonreciprocal and nonsymmetriuniform transmission lines," *IEEE Transactions on Microwave Theory and Techniques*, vol. 42, no. 2, pp. 291–297, Feb. 1994. DOI: 10.1109/22.275260 18

[4] I. V. Lindell, A. H. Sihvola, S. A. Tretyakov, and A. J. Viitanen, *Electromagnetic Waves in Chiral and Bi-Isotropic Media*. Massachusetts: Artech House, 1994. 18

[5] D. M. Pozar, *Microwave Engineering*, 3rd ed. New Jersey: Wiley, 2005. 21, 22, 23, 29

[6] O. F. Siddiqui, M. Mojahedi, and G. V. Eleftheriades, "Periodically loaded transmission line with effective negative refractive index and negative group velocity," *IEEE Transactions on Antennas and Propagation*, vol. 51, no. 10, pp. 2619–2625, Oct. 2003. DOI: 10.1109/TAP.2003.817556 21

[7] G. V. Eleftheriades and K. G. Balmain, *Negative-Refraction Metamaterials*. New Jersey: Wiley Interscience, 2005. 21

[8] N. Chamnandechakun and D. Torrungrueng, "An Analytical Study of Terminated FinitePeriodic Structures," in *Proceeding of the 27th Electrical Engineering Conference*, Nov. 2004, vol. 2, pp. 369–372. 23

2.7   PROBLEMS

2.1. Show that the voltage reflection coefficient Γ at the load and the input impedance Z_{in} of NRLUTLs as shown in Figure 2.1 are given as in (2.7) and (2.8), respectively.

2.2. Show that the phasor voltage $V(z)$ and the phasor current $I(z)$ distributed along ETL-NUTL scan be expressed explicitly in terms of the superposition of two traveling waves propagating in the opposite directions as shown in (2.12) and (2.13), respectively.

2.3. Using the *ABCD* matrix technique and eigenanalysis [5], show that the formula of the Bloch impedances $Z_B^{\pm}$ in terms of the total *ABCD* parameters of the unit cell of reciprocal lossless periodic TL structures is given as (2.18).

2.4. Determine the total *ABCD* parameters of the unsymmetrical unit cell in Figure 2.2 as a function of the angular frequency ω. Find the determinant of the total *ABCD* matrix. Is the determinant equal to unity independent of ω? Explain in detail.

2.5. Show that the total *ABCD* matrix of a *reciprocal lossless* unit cell possesses the following properties: A and D are *purely real* and B and C are *purely imaginary*.

2.6. Show that the total *ABCD* matrix of a *symmetric* unit cell possesses the following property: A and D are *identical*.

2.7. Derive (2.24) and (2.25) for the terminated reciprocal lossless periodic TL structure.

2.8. Verify that the *matching condition* for terminated periodic TL structures can be obtained when $Z_L = Z_0^+$ for the NNCR case and $Z_L = -Z_0^-$ for the NCR case.

2.9. Determine the total *ABCD* parameters of the lossless unsymmetrical unit cell in Figure 2.5 as a function of the angular frequency ω by using $d = 2$ cm, $Z_1 = 150\,\Omega$, $C_s = 5.6\,\text{pF}$, and $C_{sh} = 0.44\,\text{pF}$. It is assumed that the phase velocity of wave propagating along the uniform TLs is equal to 3×10^8 m/s. In addition, plot both magnitude and phase of each *ABCD* parameter from 0.5 to 10 GHz.

2.10. Verify the plots in Figures 2.6 to 2.9.

CHAPTER 3

Theory of Bi-Characteristic-Impedance Transmission Lines (BCITLs)

3.1 INTRODUCTION

In Chapter 2, the theory of CCITLs is developed in detail. Due to the fact that CCITLs are lossless, they cannot account for losses associated with lossy TLs and/or lossy lumped elements to obtain higher accuracy in modeling practical TL problems. Thus, one needs to resort to a more general model for these lossy elements, which is the model based on *bi-characteristic-impedance transmission lines* (BCITLs). By definition, BCITLs are generally lossy transmission lines possessing two complex characteristic impedances Z_0^+ and Z_0^- for forward and reverse waves respectively, where $Z_0^\pm$ are generally *different* and *not* complex conjugate of one another, as shown in Figure 3.1. In addition, BCITLs generally possess different complex propagation constants γ^+ and γ^- for forward and reverse waves, respectively. In Figure 3.1, the BCITL of length ℓ is terminated in a passive load impedance Z_L, where Γ is the voltage reflection coefficient at the load and Z_{in} is the input impedance of the terminated BCITL. Using the same concept as in Appendix B, it can be shown rigorously that $\gamma^+ \neq \gamma^-$ is the required condition for *nonreciprocal* BCITLs.

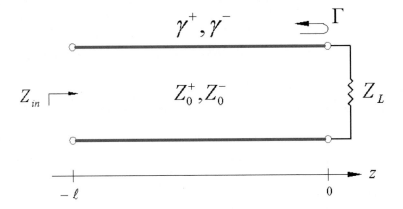

Figure 3.1: A BCITL terminated in a passive load impedance Z_L.

It should be pointed out that the CCITL is a special case of the BCITL when Z_0^+ and Z_0^- are complex conjugate of one another; i.e., when losses associated with the BCITL vanish, the BCITL is essentially the CCITL. Examples of BCITLs are RLSUTLs, nonreciprocal lossy uniform transmission lines (NRLSUTLs), finite reciprocal lossy periodic TL structures and active loaded TLs [1]–[3]. Note that RLSUTLs are discussed in detail earlier in Section 1.3.1, where their characteristic impedances Z_0 are complex. Thus, the RLSUTL is a special case of BCITLs with the identical Z_0^+ and Z_0^-; i.e., $Z_0^+ = Z_0^- = Z_0$. For active loaded TLs, active devices (e.g., microwave transistor amplifiers) are connected to transmission lines in cascade, which are not discussed in this book. Details of active loaded TLs can be found in [3]. In this chapter, only NRLSUTLs and finite reciprocal lossy periodic TL structures are discussed.

This chapter is organized as follows. Section 3.2 presents nonreciprocal lossy uniform transmission lines. Finite reciprocal lossy periodic TL structures are discussed in Section 3.3. Finally, conclusions are provided in Section 3.4.

3.2   NONRECIPROCAL LOSSY UNIFORM TRANSMISSION LINES

In Figure 3.1, the traveling wave equations for the phasor voltage $V(z)$ and the phasor current $I(z)$ along NRLSUTLs can be written as

$$V(z) = V_0^+ e^{-\gamma^+ z} + V_0^- e^{\gamma^- z}, \tag{3.1}$$

$$I(z) = \frac{V_0^+}{Z_0^+} e^{-\gamma^+ z} - \frac{V_0^-}{Z_0^-} e^{\gamma^- z}. \tag{3.2}$$

Note that V_0^+ and V_0^- are the voltage amplitudes for incident and reflected waves at the load ($z = 0$), respectively. Z_0^+ and Z_0^- are generally different, and γ^+ generally differs from γ^- as well. Thus, NRLSUTLs are *nonreciprocal* BCITLs. It is observed that these voltage and current equations for NRLSUTLs are similar to those for NRLUTLs (see (2.1) and (2.2)). Note that the exponential factors in (3.1) and (3.2) include loss effects of transmission lines, which are more general than those in (2.1) and (2.2).

For convenience in constructing a graphical tool for analysis and design of BCITLs in Chapter 4, the characteristic impedances $Z_0^{\pm}$ of NRLSUTLs are represented in the rectangular form as

$$Z_0^+ = P + jQ, \tag{3.3}$$

$$Z_0^- = U + jV, \tag{3.4}$$

where P and Q are the real and imaginary parts of the complex characteristic impedance Z_0^+ respectively, and U and V are the real and imaginary parts of the complex characteristic impedance Z_0^-, respectively. Applying the load condition and using (3.1) and (3.2), the voltage reflection coefficient at the load Γ as shown in Figure 3.1 is found to be

$$\Gamma \equiv \frac{V_0^-}{V_0^+} = \frac{Z_L Z_0^- - Z_0^+ Z_0^-}{Z_L Z_0^+ + Z_0^+ Z_0^-} \tag{3.5}$$

In addition, the input impedance Z_{in} can be computed from Γ as follows:

$$Z_{in} \equiv \frac{V(z=-\ell)}{I(z=-\ell)} = Z_0^+ Z_0^- \frac{1 + \Gamma e^{-2\tilde{\gamma}\ell}}{Z_0^- - Z_0^+ \Gamma e^{-2\tilde{\gamma}\ell}}, \tag{3.6}$$

where (3.1), (3.2), and (3.5) are employed. Note that $\tilde{\gamma}$ in (3.6) is the *arithmetic* mean of $\gamma^\pm$, which can be expressed as

$$\tilde{\gamma} = \frac{1}{2}\left(\gamma^+ + \gamma^-\right). \tag{3.7}$$

It is observed that the formulas of Γ for both NRLSUTLs and NRLUTLs are identical (see (3.5) and (2.7)). However, the formula of Z_{in} for NRLSUTLs is different from that of NRLUTLs in that the former employs the arithmetic mean of $\gamma^\pm$ while the latter uses the arithmetic mean of $\beta^\pm$ (See (2.9)) as shown in (3.6) and (2.8), respectively.

3.3   FINITE RECIPROCAL LOSSY PERIODIC TRANSMISSION LINE STRUCTURES

In Section 2.5, finite reciprocal *lossless* periodic TL structures are discussed in detail. However, periodic TL structures may contain *lossy* elements in practice; e.g., lossy TLs and lossy lumped elements. Figure 3.2 illustrates an example of an *unsymmetrical* unit cell associated with reciprocal lossy periodic TL structures. The unit cell consists of a length d of reciprocal lossless uniform transmission line loaded with lossy lumped resistors (R_S and R_{SH}) across the midpoint of the line. It is assumed that these lumped resistors have no physical length at the frequency band of interest. The characteristic impedance and the propagation constant of the *unloaded* reciprocal lossless uniform transmission line are defined as Z_1 and k, respectively.

The wave equations for the phasor voltage V_m and the phasor current I_m at the terminal of the mth unit cell as shown in Figure 3.3 (where $m = 1, 2, \ldots, M$) can be written compactly as [4]

$$V_m = V_0^+ e^{-m\gamma d} + V_0^- e^{m\gamma d}, \tag{3.8}$$

$$I_m = \frac{V_0^+}{Z_B^+} e^{-m\gamma d} + \frac{V_0^-}{Z_B^-} e^{m\gamma d}, \tag{3.9}$$

where V_0^+ and V_0^- are defined as the amplitudes of the incident and reflected voltage waves referenced at the input of the reciprocal lossy periodic TL structure, respectively. Note that V_0 and I_0 in Figure 3.3 denote the total input voltage and the total input current of the terminated periodic TL structure of M unit cells, respectively. V_0 and I_0 can be obtained from (3.8) and (3.9) by letting m

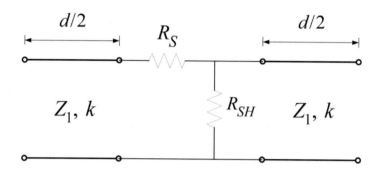

Figure 3.2: An example of an *unsymmetrical* unit cell associated with reciprocal lossy periodic TL structures.

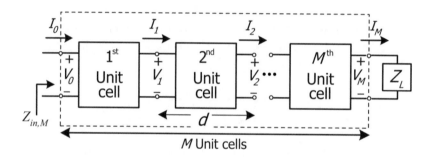

Figure 3.3: A finite reciprocal lossy periodic TL structure of M unit cells terminated in a load impedance Z_L.

= 0, respectively. In (3.9), $Z_B^{\pm}$ are referred to as the Bloch impedances, which can be expressed in terms of the *total ABCD* parameters of the unit cell as

$$Z_B^{\pm} = \frac{-2B}{A - D \mp j\sqrt{4 - (A + D)^2}},\tag{3.10}$$

which is identical to (2.20) for the lossless case. Note that Z_B^+ and Z_B^- are the Bloch impedances for forward and reverse waves, respectively. In (3.8) and (3.9), $\gamma = \alpha + j\beta$ is the *effective* complex propagation constant of the reciprocal lossy periodic TL structure, where α and β are the *effective* attenuation constant and *effective* propagation constant of the structure, respectively. Mathematically, γ can be determined from the following equation [4]

$$\cosh(\gamma d) = \frac{A + D}{2}.\tag{3.11}$$

Let us consider a finite reciprocal lossy periodic TL structure of M unit cells terminated in a load impedance Z_L as illustrated in Figure 3.3. It can be shown that the input impedance $Z_{in,M}$ at the input terminal of the terminated structure can be expressed as (see (2.24))

$$Z_{in,M} \equiv \frac{V_0}{I_0} = Z_B^+ Z_B^- \frac{1 + \Gamma e^{-2M\gamma d}}{Z_B^- + Z_B^+ \Gamma e^{-2M\gamma d}}, \tag{3.12}$$

where Γ is the voltage reflection coefficient *at the load*, which is given by

$$\Gamma = -\frac{Z_L Z_B^- - Z_B^+ Z_B^-}{Z_L Z_B^+ - Z_B^+ Z_B^-}. \tag{3.13}$$

Note that *reciprocal* BCITLs can be used to model finite reciprocal lossy periodic TL structures as shown in Figure 3.1 with $\gamma^+ = \gamma^- = \gamma$. Using (3.6) with $\gamma^+ = \gamma^- = \gamma$, the input impedance Z_{in} of the terminated reciprocal BCITL can be written as

$$Z_{in} = Z_0^+ Z_0^- \frac{1 + \Gamma e^{-2\gamma\ell}}{Z_0^- - Z_0^+ \Gamma e^{-2\gamma\ell}}, \tag{3.14}$$

where Γ is the voltage reflection coefficient *at the load* defined as in (3.5). Comparing (3.12) to (3.14), $Z_B^\pm$ and $Z_0^\pm$ are related as

$$Z_B^\pm = \pm Z_0^\pm. \tag{3.15}$$

Using (3.12) and (3.15), the input impedance $Z_{in,M}$ at the input terminal of the terminated structure in Figure 3.3 can be rewritten in terms of $Z_0^\pm$ as

$$Z_{in,M} = Z_0^+ Z_0^- \frac{1 + \Gamma e^{-2M\gamma d}}{Z_0^- - Z_0^+ \Gamma e^{-2M\gamma d}}, \tag{3.16}$$

where $Z_0^\pm$ in (3.16) are defined as the *effective* characteristic impedances of the reciprocal lossy periodic TL structure. From (3.16), it can be concluded that the equation of the input impedance of terminated finite reciprocal lossy periodic TL structures is the same as that of terminated reciprocal BCITLs. Thus, these lossy periodic TL structures *when considering at the input terminal of each unit cell* can be effectively analyzed using reciprocal BCITLs. Like CCITLs, lossy periodic TL structures are probably the most useful example of BCITLs in practice.

It is interesting to point out that reciprocal lossy periodic TL structures can exhibit both NNCRs ($\text{Re}\{Z_0^\pm\} \geq 0$) and NCRs ($\text{Re}\{Z_0^\pm\} < 0$) as shown in an example below. For reciprocal lossy periodic TL structures with NNCRs, (3.3) and (3.4) implies that $P \geq 0$ and $U \geq 0$. For those with NCRs, $P < 0$ and $U < 0$ are required. Like the CCITL case (see Section 2.5), it can be shown using (3.5) and (3.16) that the *matching condition* resulting in $Z_{in,M} = Z_L$ for reciprocal lossy periodic TL structures can be obtained when $Z_L = Z_0^+$ for the NNCR case and $Z_L = -Z_0^-$ for the NCR case; i.e., the magnitude of the voltage reflection coefficient at the load, $|\Gamma|$, is equal to zero and approaches infinity, respectively.

An example of reciprocal lossy periodic TL structures of Figure 3.2 terminated in a passive load impedance Z_L is shown in Figure 3.4, where the number of unit cells is equal to one ($M=1$) for simplicity. Note that Z_{in} is the input impedance of the terminated reciprocal lossy periodic TL structures. In this example, the parameters in Figure 3.4 are given as follows: $d = 2$ cm, $Z_1 = 50$ Ω, $R_S = 75$ Ω, $R_{SH} = 200$ Ω and $Z_L = 50$ Ω [5]. It is assumed that the phase velocity of wave propagating along the reciprocal lossless uniform TL is equal to 3×10^8 m/s. The frequency of interest ranges from 1 GHz to 10 GHz.

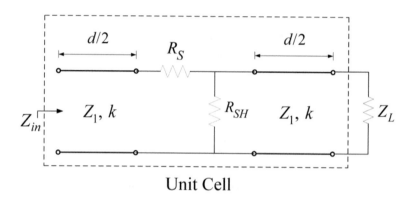

Unit Cell

Figure 3.4: An example of an *unsymmetrical* unit cell associated with reciprocal lossy periodic TL structures ($M = 1$) terminated in a passive load impedance Z_L.

Using (3.10) and (3.15), the plots of the magnitudes and the arguments of $Z_0^{\pm}$ versus frequency are shown in Figures 3.3 and 3.6, respectively. Note that ϕ^+ and ϕ^- in Figure 3.6 are the arguments of Z_0^+ and Z_0^-, respectively. From Figure 3.6, the reciprocal lossy periodic TL structure ($M \geq 1$) for the unit cell given in Figure 3.4 exhibits NNCRs ($-90° \leq \phi^{\pm} \leq 90°$) in the frequency range of 1.0 to 7.5 GHz and NCRs ($-180° < \phi^{\pm} < -90° \cup 90° < \phi^{\pm} \leq 180°$) in the frequency range of 7.5 to 10 GHz. It is observed from Figures 3.3 and 3.6 that different magnitudes and phases of complex characteristic impedances, Z_0^+ and Z_0^-, are obtained as expected.

Using (3.11) to determine the effective complex propagation constant $\gamma = \alpha + j\beta$, it is observed that α is positive and negative for NNCRs and NCRs, respectively, as shown in Figure 3.7. Using the method based on short-circuited and open-circuited terminations [5], [6], it can be shown rigorously that α of BCITLs, implemented using finite reciprocal lossy periodic TL structures, is always *nonnegative* and *nonpositive* for NNCRs and NCRs, respectively. In addition, Figure 3.7 shows that β is positive in the frequency range of interest for both NNCRs and NCRs. Furthermore, Figure 3.8 illustrates the plot of the magnitude of Γ versus frequency. From the plot, it is observed that $|\Gamma| \leq 1$ and $|\Gamma| \geq 1$ for NNCRs and NCRs in this example, respectively. However, it should be pointed out that this observation for the magnitude of Γ is *not* valid for general BCITLs and general passive load impedances.

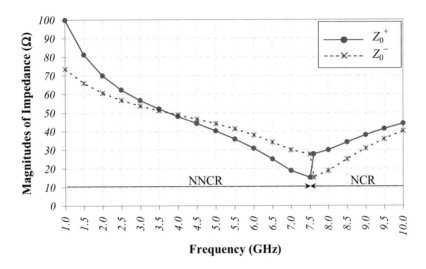

Figure 3.5: The plot of the magnitudes of Z_0^+ and Z_0^- versus frequency.

3.4   CONCLUSIONS

In this chapter, the theory of BCITLs is discussed in detail. Practical examples of BCITLs are also provided, and lossy periodic TL structures are probably the most useful example of BCITLs. Like CCITLs, it is interestingly found that BCITLs can exhibit nonnegative characteristic resistances ($\text{Re}\{Z_0^\pm\} \geq 0$) and negative characteristic resistances ($\text{Re}\{Z_0^\pm\} < 0$). In addition, the *effective* attenuation constant α of BCITLs, implemented using finite reciprocal lossy periodic TL structures, is always nonnegative and nonpositive for NNCRs and NCRs, respectively. To gain more physical understanding and simplify the analysis and design of BCITLs, graphical tools based on the Smith chart will be developed in Chapter 4.

BIBLIOGRAPHY

[1] D. Torrungrueng, P.Y. Chou, and M. Krairiksh, "A graphical tool for analysis and designof bi-characteristic-impedance transmission lines," *Microwave and OpticalTechnologyLetters*, vol. 49, no. 10, pp. 2368–2372, Oct. 2007. DOI: 10.1002/mop.22801 32

[2] D. Torrungrueng, P.Y. Chou, and M. Krairiksh, "Erratum:A graphical tool for analysis and design of bi-characteristic-impedance transmission lines," *Microwave and OpticalTechnologyLetters*, vol. 51, no. 4, pp. 1154, Apr. 2009. DOI: 10.1002/mop.22801

[3] C. Lertsirimitand D. Torrungrueng, "Analysis of active loaded transmission line usingan equivalent BCITL model," in *2007 Asia-Pacific Microwave Conference Proceedings*, Dec. 2007, vol. 4, pp. 2353–2356. DOI: 10.1109/APMC.2007.4554830 32

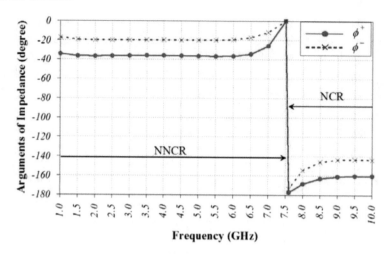

Figure 3.6: The plot of ϕ^+ and ϕ^- versus frequency.

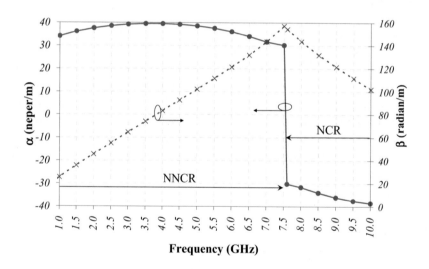

Figure 3.7: The plot of α and β versus frequency.

[4] D. M. Pozar, *Microwave Engineering*, 3rd ed. New Jersey: Wiley, 2005. 33, 34

[5] S. Lamultree, D. Torrungrueng and C. Phongcharoenpanich, "Measurement of BCITL parameters," in *Proceedings of the 2008 ECTI International Conference*, May 2008, vol. 1, pp. 257–260. DOI: 10.1109/ECTICON.2008.4600421 36

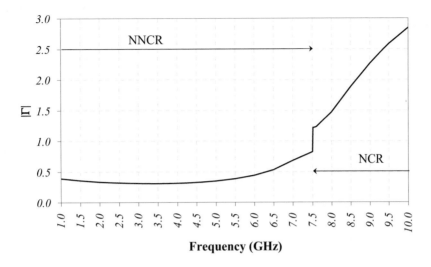

Figure 3.8: The plot of the magnitude of Γ versus frequency.

[6] D. Torrungrueng, S. Lamultree, C. Phongcharoenpanich and M. Krairiksh, "In-depth analysis of reciprocal periodic structures of transmission lines," *IET Microwaves, Antennas & Propagation*, vol. 3, no. 4, pp. 591–600, Jun. 2009. DOI: 10.1049/iet-map.2008.0205 36

3.5   PROBLEMS

3.1. Show that the input impedance Z_{in} of NRLSUTLs as shown in Figure 3.1 is given as in (3.6).

3.2. Determine the total *ABCD* parameters of the lossy unsymmetrical unit cell in Figure 3.4 as a function of the angular frequency ω by using $d = 2$ cm, $Z_1 = 50$ Ω, $R_S = 75$ Ω and $R_{SH} = 200$ Ω. It is assumed that the phase velocity of wave propagating along the uniform TL is equal to 3×10^8 m/s. In addition, plot both magnitude and phase of each *ABCD* parameter from 1 to 10 GHz.

3.3. Verify the plots in Figures 3.5 to 3.8.

CHAPTER 4

Meta-Smith Charts for CCITLs and BCITLs

4.1   INTRODUCTION

Analytical approaches shown in Chapters 2 and 3 for solving problems associated with CCITLs and BCITLs usually provide quite complex formulae. In order to assist in the analysis and design of CCITLs and BCITLs, the need for graphical tools is obvious. Like the Smith chart, graphical tools usually provide a useful way with more physical insight of visualizing associated TL phenomenon and solving related problems effectively. In this chapter, the Smith chart is generalized to problems associated with CCITLs and BCITLs, called *Meta-Smith charts*. Note that the Greek word *meta* means *beyond*. Meta-Smith charts can be applied to more practical and useful TL problems; i.e., CCITLs and BCITLs, *beyond* the capability of the Smith chart. The Meta-Smith charts are constructed using the same approach as in constructing the Smith chart. In this chapter, the Meta-Smith charts for CCITLs with NNCRs and NCRs are developed, including the Meta-Smith chart for BCITLs. Only passive loads ($\text{Re}\{Z_L\} \geq 0$) are of interest, where Z_L is the load impedance.

In the references [1]–[7], the authors employed the terminology "T-Charts" for graphical tools associated with CCITLs and BCITLs. In this book, the terminology "Meta-Smith charts" is used instead of "T-Charts" to make readers, who are familiar with the Smith chart, see the originality and generality of the proposed graphical tools, Meta-Smith charts. In addition, it should be pointed out that in the literature [2], [8] the terminologies "*passive* characteristic impedances" and "*active* characteristic impedances" refer to NNCRs and NCRs, respectively. In this book, NNCRs and NCRs are employed instead to avoid any misleading of those terminologies, especially for *active* characteristic impedances.

For convenience in derivation later, let $Z_0^{\pm}$ be the characteristic impedances of both CCITLs and BCITLs. For CCITLs, Z_0^+ and Z_0^- are *complex conjugate* of one another. For BCITLs, Z_0^+ and Z_0^- are generally *different* and *not* complex conjugate of one another. However, it is found in Chapters 2 and 3 that the voltage reflection coefficient at the load Γ of both CCITLs and BCITLs is given compactly in terms of Z_L and $Z_0^{\pm}$ as (see (2.7) and (3.5))

$$\Gamma = \frac{Z_L Z_0^- - Z_0^+ Z_0^-}{Z_L Z_0^+ + Z_0^+ Z_0^-}. \tag{4.1}$$

Like the Smith chart, each Meta-Smith chart is essentially a polar plot of the voltage reflection coefficient at the load Γ of CCITLs or BCITLs. It can be employed to convert from voltage reflection coefficients to *normalized* impedances (or admittances), and vice versa. For convenience in studying CCITLs and BCITLs, computerized Meta-Smith chart programs are developed, and they are employed in generating Meta-Smith charts in this book.

This chapter is organized as follows. The Meta-Smith chart for CCITLs with NNCRs is discussed in detail in Section 4.2, and Section 4.3 presents the Meta-Smith chart for CCITLs with NCRs. The Meta-Smith chart for BCITLs with NNCRs and NCRs is illustrated in Section 4.4. Finally, conclusions are provided in Section 4.5.

4.2   THE META-SMITH CHART FOR CCITLS WITH NONNEGATIVE CHARACTERISTIC RESISTANCES

The Meta-Smith chart for CCITLs with NNCRs ($\text{Re}\{Z_L\} \geq 0$) [1], [9] is constructed by using the same concept as in constructing the standard ZY Smith chart in [10] (see Problem 1.3 at the end of Chapter 1 as well), and it is discussed in detail in this section. Note that this Meta-Smith chart is a superposition of the *Z Meta-Smith chart* and the *Y Meta-Smith chart*. The Z Meta-Smith chart consists of resistance and reactance circles in the Γ plane, while the Y one is a superposition of conductance and susceptance circles in the Γ plane.

For convenience, let us define $Z_0^{\pm}$ in the polar form as

$$Z_0^{\pm} = \left| Z_0^{\pm} \right| e^{\mp j\phi}, \tag{4.2}$$

where $|Z_0^{\pm}|$ and ϕ are the absolute value and the argument of Z_0^{-}, respectively. As pointed out earlier in Chapter 2, the argument ϕ for CCITLs with NNCRs must lie in the following range:

$$-90° \leq \phi \leq 90°. \tag{4.3}$$

Due to the fact that the magnitude of the voltage reflection coefficient at the load $|\Gamma|$ is always less than or equal to unity for terminated CCITLs with NNCRs, the region of interest is always *inside or on* the unit circle in the Γ plane for *passive* load terminations (see Appendix D).

4.2.1   THE Z META-SMITH CHART FOR CCITLS WITH NNCRS

For CCITLs, the normalized impedance z is defined as [1]

$$z \equiv \frac{Z}{\tilde{Z}_0}, \tag{4.4}$$

where $\tilde{Z}_0$ is the *geometric mean* of the characteristic impedances $Z_0^{\pm}$; i.e.,

$$\tilde{Z}_0 = \sqrt{Z_0^+ Z_0^-} = \left| Z_0^{\pm} \right|. \tag{4.5}$$

From (4.5), it should be noted that $\tilde{Z}_0$ is always *real* and *positive* for physically realizable CCITLs. Using (4.1) and (4.4), Γ can be rewritten compactly as

$$\Gamma = \frac{z_L z_0^- - 1}{z_L z_0^+ + 1}, \tag{4.6}$$

where z_L is the normalized load impedance and $z_0^\pm$ are the normalized characteristic impedances. Using (4.2), (4.4), and (4.5), $z_0^\pm$ are found to be

$$z_0^\pm = e^{\mp j\phi}. \tag{4.7}$$

Note that $z_0^\pm$ in (4.7) depend only on the argument ϕ only. To derive equations for resistance and reactance circles, it is better to express z_L explicitly in terms of Γ. Using (4.6), z_L can be expressed compactly as

$$z_L = \frac{1 + \Gamma}{z_0^- - z_0^+ \Gamma}. \tag{4.8}$$

For convenience in manipulation, let us express Γ and z_L as $\Gamma = \Gamma_r + j\Gamma_i$ and $z_L = r_L + jx_L$, where Γ_r and Γ_i are the real and imaginary parts of Γ respectively, and r_L and x_L are the real and imaginary parts of z_L, respectively. Substituting these expressions of Γ and z_L into (4.8) and using (4.7) yield

$$R_L + jx_L = \frac{1 + \Gamma_r + j\Gamma_i}{e^{j\phi} - e^{-j\phi}(\Gamma_r + j\Gamma_i)}. \tag{4.9}$$

After some straightforward algebraic manipulation, r_L and x_L in (4.9) can be expressed in terms of Γ_r, Γ_i, and ϕ as follows:

$$r_L = \frac{\cos\phi - \Gamma_r^2 \cos\phi - \Gamma_i^2 \cos\phi}{1 - 2\Gamma_r \cos 2\phi - 4\Gamma_i \cos\phi \sin\phi + \Gamma_r^2 + \Gamma_i^2}, \tag{4.10}$$

$$x_L = \frac{-\sin\phi - 2\Gamma_r \sin\phi + 2\Gamma_i \cos\phi - \Gamma_r^2 \sin\phi - \Gamma_i^2 \sin\phi}{1 - 2\Gamma_r \cos 2\phi - 4\Gamma_i \cos\phi \sin\phi + \Gamma_r^2 + \Gamma_i^2}. \tag{4.11}$$

Rearranging (4.10) and (4.11) appropriately, the following circle equations are obtained [1]:

$$\left(\Gamma_r - \frac{r_L \cos 2\phi}{r_L + \cos\phi}\right)^2 + \left(\Gamma_i - \frac{r_L \sin 2\phi}{r_L + \cos\phi}\right)^2 = \left(\frac{\cos\phi}{r_L + \cos\phi}\right)^2, \tag{4.12}$$

$$\left(\Gamma_r - \frac{(x_L \cos 2\phi - \sin\phi)}{x_L + \sin\phi}\right)^2 + \left(\Gamma_i - \frac{(x_L \sin 2\phi + \cos\phi)}{x_L + \sin\phi}\right)^2 = \left(\frac{\cos\phi}{x_L + \sin\phi}\right)^2, \tag{4.13}$$

for resistance and reactance circles, respectively.

4.2.2 THE Y META-SMITH CHART FOR CCITLS WITH NNCRS

To construct the Y Meta-Smith Chart, the normalized admittance y must be considered instead of the normalized impedance z, where y is defined as

$$y \equiv \frac{1}{z}. \tag{4.14}$$

Let $y_L = g_L + jb_L$ be the normalized load admittance, where g_L and b_L are the normalized load conductance and susceptance, respectively. Using the same technique as in constructing the Z Meta-Smith Chart above, the equations for conductance and susceptance circles (see (4.15) and (4.16), respectively) are obtained as follows [1]:

$$\left(\Gamma_r + \frac{g_L}{g_L + \cos \phi} \right)^2 + \Gamma_i^2 = \left(\frac{\cos \phi}{g_L + \cos \phi} \right)^2, \tag{4.15}$$

$$(\Gamma_r + 1)^2 + \left(\Gamma_i + \frac{\cos \phi}{b_L - \sin \phi} \right)^2 = \left(\frac{\cos \phi}{b_L - \sin \phi} \right)^2. \tag{4.16}$$

Note that these four circle equations (see (4.12), (4.13), (4.15), and (4.16)) depend on the argument ϕ of $Z_0^{\pm}$ in a complicated fashion. It is interesting to point out that these circle equations reduce to the corresponding circle equations for the standard ZY Smith chart when $\phi = 0°$ as expected (see Problem 1.3 at the end of Chapter 1) i.e., the standard ZY Smith chart is a special case of the Meta-Smith chart for CCITLs with NNCRs. After deriving these equations, the Meta-Smith chart for CCITLs with NNCRs can be constructed readily. In the next subsection, numerical results of the Meta-Smith chart will be illustrated.

4.2.3 NUMERICAL RESULTS

From Sections 4.2.1 and 4.2.2, it is found that the Meta-Smith chart for CCITLs with NNCRs depends on the argument ϕ. To illustrate this dependence, let us consider the Meta-Smith chart when $\phi = 30°$. Figure 4.1 shows the plot of resistance circles in the Γ plane for $\phi = 30°$. Note that each circle corresponds to different values of normalized load resistances r_L. The largest circle occurs when $r_L = 0$. As r_L increases, the radius of resistance circle becomes smaller as shown in Figure 4.1. The resistance circles between the $r_L = 0$ circle and the $r_L = 1$ circle correspond to the circles with $0 < r_L < 1$. In addition, the resistance circles inside the $r_L = 1$ circle correspond to the circles with $r_L > 1$.

Define θ as the angle between the horizontal line and the line $\overline{OP}$ drawn from the origin O in the Γ plane to the *touching point* P of all resistance circles as shown in Figure 4.1. Using (4.12), it can be shown rigorously that the relationship between the argument ϕ and the angle θ is given by

$$\theta = 2\phi. \tag{4.17}$$

In this case, θ is equal to $60°$ for $\phi = 30°$ as shown in Figure 4.1. It should be emphasized that these resistance circles of the Meta-Smith chart cannot be obtained by simply rotating the corresponding

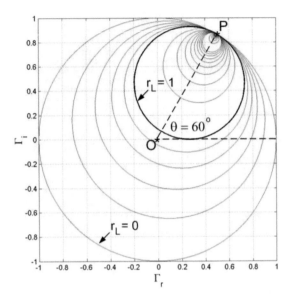

Figure 4.1: Plot of resistance circles in the Γ plane for $\phi = 30°$.

circles of the standard Smith chart by the angle θ in the *counterclockwise* direction. This is due to the fact that the *scale* of each resistance circle depends on the argument ϕ as well.

Figure 4.2 illustrates the plot of reactance circles in the Γ plane for $\phi = 30°$. From the figure, the horizontal line and the line $\overline{OP}$ drawn from the origin O in the Γ plane to the touching point P of reactance circles intersect at the angle θ defined in (4.17). Note that the point P in Figures 4.4 and 4.2 is the same point. Unlike the standard Smith chart, the reactance circle $x_L = 0$ for the Meta-Smith chart is a *curve* (i.e., a part of circle) instead of a horizontal line as in the standard Smith chart [10]. Inside the unit circle in Figure 4.2, all reactance circles above the circle $x_L = 0$ correspond to $x_L > 0$, and those below the circle $x_L = 0$ correspond to $x_L < 0$.

The Y Meta-Smith chart consists of conductance and susceptance circles as shown in Figures 4.3 and 4.4 for $\phi = 30°$, respectively. Interestingly, the touching point of conductance circles in Figure 4.3 is at the same position as in the Y Smith Chart (i.e., at the point $(-1,0)$ in the Γ plane) [10]. It is found that this point will not change when the argument ϕ changes. Each circle corresponds to different values of normalized load conductances g_L. The largest circle, which is the unit circle, occurs when $g_L = 0$. As g_L increases, the radius of conductance circle becomes smaller as shown in Figure 4.3. The conductance circles between the $g_L = 0$ circle and the $g_L = 1$ circle correspond to the circles with $0 < g_L < 1$. In addition, the conductance circles inside the $g_L = 1$ circle correspond to the circles with $g_L > 1$.

The plot of susceptance circles in the Γ plane for $\phi = 30°$ is illustrated in Figure 4.4. Like reactance circles in Figure 4.2, the susceptance circle $b_L = 0$ for the Y Meta-Smith chart as shown in

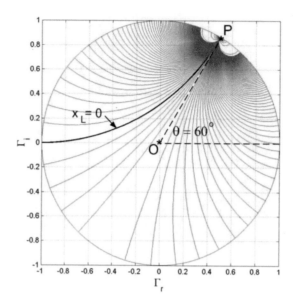

Figure 4.2: Plot of reactance circles in the Γ plane for $\phi = 30°$.

Figure 4.4 is also a *curve* (i.e., a part of circle) instead of a straight line as in the case of the Y Smith chart [10]. Inside the unit circle in Figure 4.4, all susceptance circles above and below the circle $b_L = 0$ correspond to $b_L < 0$ and $b_L > 0$, respectively. As pointed out earlier, the Y Meta-Smith chart is a superposition of conductance and susceptance circles in the Γ plane. Note that the Y Meta-Smith chart looks similar to the Y Smith chart, *but its admittance scale is totally different.*

The ZY Meta-Smith chart for CCITLs with NNCRs can be obtained by superimposing these four families of circles in Figures 4.1 to 4.4 in the Γ plane as shown in Figure 4.5 for $\phi = 30°$ as an example. From the figure, it is observed that the circle $x_L = 0$ and the circle $b_L = 0$ are identical. This can be verified by comparing (4.13) and (4.16) when $x_L = 0$ and $b_L = 0$, respectively. It is interesting to note that the circle $x_L = 0$ joins the touching point of resistance (or reactance) circles (i.e., the *open-circuited* (OC) point) and the touching point of conductance (or susceptance) circles (i.e., the *short-circuited* (SC) point) as shown in Figure 4.5. In addition, it should be pointed out that the procedures of using the ZY Meta-Smith chart in determining the input impedance of terminated CCITLs with NNCRs are *similar* to those for reciprocal lossless uniform transmission lines using the Smith chart provided in [10], [11], and they will be discussed in detail in Chapter 5.

It should be pointed out that the ZY Meta-Smith chart for CCITLs with NNCRs strongly depends on the definition of the normalized impedance defined in (4.4). Using different $\tilde{Z}_0$ results in different graphical representations of the ZY Meta-Smith chart for CCITLs with NNCRs [9]. For example, if the *arithmetic mean* of the characteristic impedances $Z_0^{\pm}$ is employed as $\tilde{Z}_0$ instead

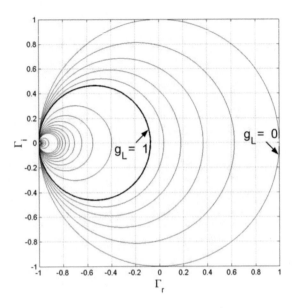

Figure 4.3: Plot of conductance circles in the Γ plane for $\phi = 30°$.

of the geometric mean of $Z_0^{\pm}$ (see (4.5)); i.e.,

$$\tilde{Z}_0 = \frac{1}{2}\left(Z_0^+ + Z_0^-\right) = \left|Z_0^{\pm}\right|\cos\phi, \tag{4.18}$$

the new four circle equations of resistance, reactance, conductance and susceptance can be obtained as follows [9]:

$$\left(\Gamma_r - \frac{r_L\cos 2\phi}{r_L + 1}\right)^2 + \left(\Gamma_i - \frac{r_L\sin 2\phi}{r_L + 1}\right)^2 = \left(\frac{1}{r_L + 1}\right)^2, \tag{4.19}$$

$$\left(\Gamma_r - \frac{(x_L\cos 2\phi - \tan\phi)}{x_L + \tan\phi}\right)^2 + \left(\Gamma_i - \frac{(x_L\sin 2\phi + 1)}{x_L + \tan\phi}\right)^2 = \left(\frac{1}{x_L + \tan\phi}\right)^2, \tag{4.20}$$

$$\left(\Gamma_r + \frac{g_L}{g_L + \cos^2\phi}\right)^2 + \Gamma_i^2 = \left(\frac{\cos^2\phi}{g_L + \cos^2\phi}\right)^2, \tag{4.21}$$

$$(\Gamma_r + 1)^2 + \left(\Gamma_i + \frac{\cos^2\phi}{b_L - \sin\phi\,\cos\phi}\right)^2 = \left(\frac{\cos^2\phi}{b_L - \sin\phi\,\cos\phi}\right)^2, \tag{4.22}$$

respectively. Comparing these four circle equations with those of using the geometric mean of $Z_0^{\pm}$ as $\tilde{Z}_0$ (see (4.12), (4.13), (4.15), and (4.16)), it is found that the corresponding equations are different indeed. However, different ZY Meta-Smith charts must yield the same results; e.g., calculated

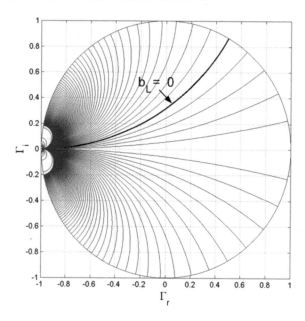

Figure 4.4: Plot of susceptance circles in the Γ plane for $\phi = 30°$.

input impedances of terminated CCITLs with NNCRs. Thus, it can be concluded that the ZY Meta-Smith chart for CCITLs with NNCRs is *not unique*. More details of nonuniqueness of ZY Meta-Smith charts can be found in [9].

4.3   THE META-SMITH CHART FOR CCITLS WITH NEGATIVE CHARACTERISTIC RESISTANCES

As discussed in Chapter 2, some CCITLs can exhibit negative characteristic resistances ($\text{Re}\{Z_0^\pm\} < 0$); e.g., finite reciprocal lossless periodic TL structures operated in passbands. For this case, the magnitude of the associated voltage reflection coefficient is always greater than or equal to unity for passive load terminations as shown in Section 2.5 (see Appendix D). Due to this reason, some modifications to the ZY Meta-Smith chart for CCITLs with NNCRs developed in Section 4.2 are necessary. The modified chart is called *the ZY Meta-Smith chart for CCITLs with NCRs*, where the range of plotting is *always outside or on* the unit circle of the voltage reflection coefficient Γ plane for passive load terminations [2].

As pointed out in Chapter 2, the argument ϕ of Z_0^- for CCITLs with NCRs must lie in the following ranges:

$$-180° < \phi < -90° \cup 90° < \phi \le 180°. \tag{4.23}$$

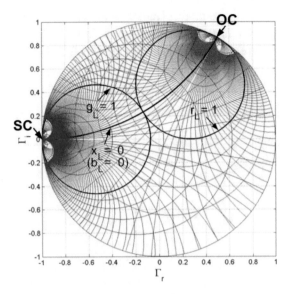

Figure 4.5: Plot of the ZY Meta-Smith chart in the Γ plane for $\phi = 30°$.

The ZY Meta-Smith chart for CCITLs with NCRs can be constructed by using the same concept as in constructing the ZY Meta-Smith chart for CCITLs with NNCRs. The ZY Meta-Smith chart for CCITLs with NCRs is also a superposition of the impedance (Z) and admittance (Y) Meta-Smith charts as discussed below.

4.3.1 THE Z AND Y META-SMITH CHARTS FOR CCITLS WITH NCRS

Due to the fact that the formulas of the voltage reflection coefficient at the load Γ of CCITLs with NNCRs and NCRs are identical (see (4.1)), all circle equations discussed in Section 4.2 (see (4.12), (4.13), (4.15), and (4.16)) remain the same for CCITLs with NCRs. However, these circle contours for CCITLs with NCRs are only considered *outside* or *on* the unit circle with the valid ranges of the argument ϕ defined in (4.23), for passive load terminations [2]. In the next subsection, numerical results of the ZY Meta-Smith chart for CCITLs with NCRs are illustrated.

4.3.2 NUMERICAL RESULTS

From the previous subsection, it is found that the ZY Meta-Smith chart for CCITLs with NCRs also depends on the argument ϕ. To illustrate this dependence, consider the case when $\phi = 150°$, which is in the ranges defined in (4.23). Figure 4.6 illustrates the plot of resistance circles in the Γ plane. Each circle corresponds to different values of normalized load resistances r_L. The boundary circle, separating the regions of passive ($r_L \geq 0$) and active ($r_L < 0$) loads, occurs when the r_L equals to zero, which is the unit circle. Note that the region of active loads is within the unit circle. As r_L increases, the radius of the resistance circle becomes larger until it reaches a "turning point," where from this point on, the radius of the circle decreases, as shown in Figure 4.6. From (4.12), this point corresponds to

$$r_L = -\cos\phi, \tag{4.24}$$

which makes the denominators in all associated terms of (4.12) equal to zero.

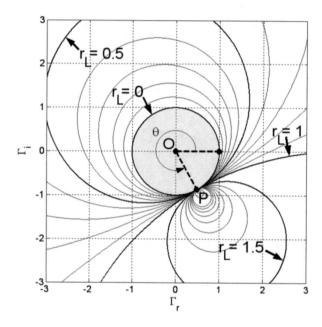

Figure 4.6: Plot of resistance circles in the Γ plane for $\phi = 150°$.

Define θ as the angle between the horizontal line and the line $\overline{OP}$, drawn from the origin O in the Γ plane to the touching point P of all resistance circles, in the counterclockwise direction as shown in Figure 4.6. The relationship between the argument ϕ and the angle θ are given by

$$\theta = 2\phi, \tag{4.25}$$

which θ is equal to 300° for $\phi = 150°$. Note that the relationship between ϕ and θ in (4.25) is identical to (4.17) for the Meta-Smith chart for CCITLs with NNCRs.

Figure 4.7 illustrates the plot of reactance circles in the Γ plane for $\phi = 150°$. From this figure, the horizontal line and the line $\overline{OP}$, drawn from the origin O to the intersecting point P of all reactance circles, forms the angle θ (in counterclockwise direction) as well. Note that the point P in Figures 4.4 and 4.7 is the same point. In addition, it is observed that the reactance circle $x_L = 0$ for the Z Meta-Smith chart for CCITLs with NCRs is a part of circle. Furthermore, all reactance circles outside the circle $x_L = 0$ correspond to $x_L < 0$, and those inside the circle $x_L = 0$ correspond to $x_L > 0$ for $90° < \phi \leq 180°$, and vice versa for $-180° < \phi < -90°$.

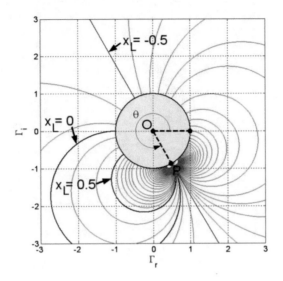

Figure 4.7: Plot of reactance circles in the Γ plane for $\phi = 150°$.

The Y Meta-Smith chart for CCITLs with NCRs consists of conductance circles and susceptance circles as shown in Figure 4.8 for $\phi = 150°$. It can be observed that these circles intersect (or touch) at the same point on the chart, which is the point $(-1, 0)$ in the Γ plane. It is interesting to point out that this point remains the same when the argument ϕ changes. For conductance circles, the boundary circle separating the regions of passive and active loads occurs when $g_L = 0$, which is the unit circle. As g_L increases to the turning point which corresponds to

$$g_L = -\cos\phi, \tag{4.26}$$

the radius of the conductance circle becomes larger, while after that, the radius becomes smaller. Unlike the reactance circles in Figure 4.7, the susceptance circles inside and outside the circle $b_L = 0$ corresponds to $b_L < 0$ and $b_L > 0$ for $90° < \phi \leq 180°$ respectively, and vice versa for $-180° < \phi < -90°$.

Finally, the ZY Meta-Smith chart for CCITLs with NCRs can be obtained by superimposing these four families of circles in Figures 4.6 to 4.8 in the Γ plane, as shown in Figure 4.9 for $\phi = 150°$. From the figure, it is observed that the circle $x_L = 0$ and the circle $b_L = 0$ are the same. This can be verified by comparing (4.13) and (4.16) where $x_L = 0$ and $b_L = 0$, respectively. It is interesting to note that the circle $x_L = 0$ joins the intersecting (or touching) point of the Z Meta-Smith chart for CCITLs with NCRs (the point P in Figures 4.6 and 4.7) and the intersecting (or touching) point of the Y one (the point $(-1, 0)$), as shown in Figure 4.9. Furthermore, the former and latter intersecting

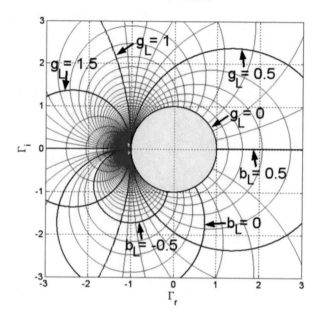

Figure 4.8: Plot of the Y Meta-Smith chart for CCITLs with NCRs in the Γ plane for $\phi = 150°$.

(or touching) points are the OC and SC points, respectively. In addition, it should be pointed out that the resistance and reactance circles intersect at the right angle due to the conformal mapping property [12]. Similarly, the conductance and susceptance circles intersect at the right angle as well. To illustrate one more example of the ZY Meta-Smith chart for CCITLs with NCRs, Figure 4.10 shows the ZY Meta-Smith chart in the Γ plane for $\phi = -150°$. Note that the circles $r_L = 1$ and x_L = 0 are reversed to the opposite side of the horizontal axis in the Γ plane compared to the previous case (see Figure 4.9). For other values of ϕ defined in (4.23), all observations above for $\phi = \pm150°$ can be applied as well.

   It should be pointed out that the procedures of using the ZY Meta-Smith chart for CCITLs with NCRs in determining the input impedance of terminated CCITLs are *similar* to those for reciprocal lossless uniform transmission lines using the Smith chart, and they will be discussed in detail in Chapter 5. However, the operating region of the ZY Meta-Smith chart for CCITLs with NCRs is *outside* or *on* the unit circle in the Γ plane for passive load terminations, which is different from that of the Smith chart.

   For more convenience in usage, the ZY Meta-Smith chart for CCITLs with NCRs can be modified appropriately, called *a modified ZY Meta-Smith chart for CCITLs with NCRs* (originally called a modified extended ZY T-chart [13], [14]), by mapping the region outside the unit circle of the former to the inside one of the latter using an appropriate *bilinear transformation* [12]. The modified chart is plotted in the plane of *the inverse of the voltage reflection coefficient* instead of the

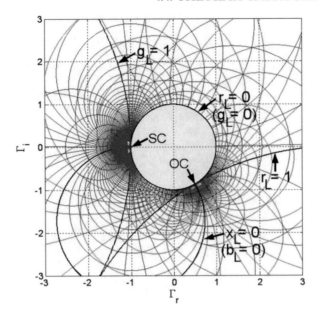

Figure 4.9: Plot of the ZY Meta-Smith chart for CCITLs with NCRs in the Γ plane for $\phi = 150°$.

Γ plane. It is found that the modified chart also depends on the argument ϕ defined in (4.23) in a complicated fashion.

4.4   THE META-SMITH CHART FOR BCITLS

In this section, the Meta-Smith charts for CCITLs with NNCRs and NCRs are generalized to treat BCITL problems, called *the Meta-Smith chart for BCITLs* [3], [4]. Due to the fact that $Z_0^{\pm}$ of BCITLs are generally not complex conjugate of one another, associated equations for the Meta-Smith chart for BCITLs will be found to be more complicated compared to those for the Meta-Smith charts for CCITLs with NNCRs and NCRs. As pointed out in Chapter 3, BCITLs can exhibit both NNCRs and NCRs, which will be treated simultaneously.

The Meta-Smith chart for BCITLs can be constructed by using the same concept as in constructing the Meta-Smith charts for CCITLs developed in Sections 4.2 and 4.3. Using (4.1), it can be shown that the magnitude of the voltage reflection coefficient at the load $|\Gamma|$ associated with terminated BCITLs may be greater than unity for some passive load impedances Z_L. This implies that associated circles are not only plotted inside the unit circle *but as well as outside* of the unit circle in the complex Γ plane. As pointed out in Section 1.3.1, reciprocal lossy uniform TLs also exhibit this phenomenon; i.e., their magnitude of the voltage reflection coefficient at the load could be greater than unity for some passive load terminations. As pointed out in Chapter 3 , a reciprocal

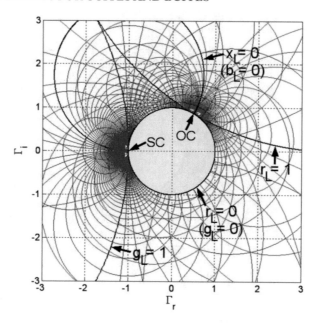

Figure 4.10: Plot of the ZY Meta-Smith chart for CCITLs with NCRs in the Γ plane for $\phi = -150°$.

lossy uniform TL is one type of BCITLs. Thus, it is not surprised to see the same phenomenon discussed above for terminated BCITLs.

Note that the Meta-Smith chart for BCITLs is a superposition of the *Z Meta-Smith chart* and the *Y Meta-Smith chart*. Next, the key steps to obtain associated circle equations of the Meta-Smith chart for BCITLs are discussed below.

4.4.1 THE Z META-SMITH CHART FOR BCITLS

For BCITLs, the normalized impedance z of any impedance Z is defined as in the case of CCITLs (see (4.4) and (4.5)); i.e.,

$$z = \frac{Z}{\sqrt{Z_0^+ Z_0^-}}, \tag{4.27}$$

where $Z_0^{\pm}$ are defined in the rectangular form as in (3.3) and (3.4). Unlike CCITLs, the term $\sqrt{Z_0^+ Z_0^-}$ in (4.27) is *complex* in general. Using (4.41) for the voltage reflection coefficient at the load for BCITLs and (4.27), the normalized load impedance z_L can be expressed compactly in terms of the normalized characteristic impedances $z_0^{\pm}$ and Γ as

$$z_L = \frac{1 + \Gamma}{z_0^- - z_0^+ \Gamma}, \tag{4.28}$$

where z_0^+ and z_0^- are defined as

$$z_0^+ \equiv \frac{Z_0^+}{\sqrt{Z_0^+ Z_0^-}} = \frac{P + jQ}{\sqrt{Z_0^+ Z_0^-}}, \tag{4.29}$$

$$z_0^- \equiv \frac{Z_0^-}{\sqrt{Z_0^+ Z_0^-}} = \frac{U + jV}{\sqrt{Z_0^+ Z_0^-}}. \tag{4.30}$$

For convenience in manipulation, let $\Gamma = \Gamma_r + j\Gamma_i$ and $z_L = r_L + jx_L$. Substituting Γ and z_L into (4.28) and considering only the real and imaginary parts of (4.28), with some straightforward manipulations, the following resistance and reactance circle equations are obtained:

$$\left(\Gamma_r + \frac{A}{2}\right)^2 + \left(\Gamma_i + \frac{B}{2}\right)^2 = \left(\frac{A}{2}\right)^2 + \left(\frac{B}{2}\right)^2 - C, \tag{4.31}$$

$$\left(\Gamma_r + \frac{D}{2}\right)^2 + \left(\Gamma_i + \frac{E}{2}\right)^2 = \left(\frac{D}{2}\right)^2 + \left(\frac{E}{2}\right)^2 - F, \tag{4.32}$$

respectively. Note that A, B, and C in (4.31) are dependent on the normalized load resistance r_L and the normalized characteristic impedance parameters, p, q, u, and v, which are defined as follows:

$$A = -\frac{[2r_L (up + vq) + (u - p)]}{r_L (p^2 + q^2) + p}, \tag{4.33}$$

$$B = \frac{[2r_L (uq - vp) - (v + q)]}{r_L (p^2 + q^2) + p}, \tag{4.34}$$

$$C = \frac{r_L (u^2 + v^2) - u}{r_L (p^2 + q^2) + p}. \tag{4.35}$$

Note that p and q are the real and imaginary parts of z_0^+ (see (4.29)) respectively, and u and v are the real and imaginary parts of z_0^- (see (4.30)), respectively. In (4.32), D, E, and F are dependent on the normalized load reactance x_L and the parameters p, q, u, and v, which are defined as

$$D = -\frac{[2x_L (up + vq) + (q - v)]}{x_L (p^2 + q^2) - q}, \tag{4.36}$$

$$E = \frac{[2x_L (uq - vp) - (p + u)]}{x_L (p^2 + q^2) - q}, \tag{4.37}$$

$$F = \frac{x_L (u^2 + v^2) + v}{x_L (p^2 + q^2) - q}. \tag{4.38}$$

It can be shown that (4.31) and (4.32) can be reduced to the equations of the resistance and reactance circles of the Z Meta-Smith charts for CCITLs (see (4.12) and (4.13)), respectively, when $p = u$ and $q = -v$, (i.e., $Z_0^+ = (Z_0^-)^*$). Thus, the Z Meta-Smith charts for CCITLs are special cases of the Z Meta-Smith chart for BCITLs when $Z_0^+ = (Z_0^-)^*$ indeed.

4.4.2   THE Y META-SMITH CHART FOR BCITLS

To construct the Y Meta-Smith chart for BCITLs, the normalized load admittance y_L must be considered instead, where y_L is the reciprocal of z_L. Let $y_L = g_L + jb_L$ be the normalized load admittance as defined in Section 4.2.2. Following the same procedures as in constructing the Z Meta-Smith chart for BCITLs, the equations for the conductance and susceptance circles are obtained, respectively, as follows:

$$\left(\Gamma_r + \frac{G}{2}\right)^2 + \left(\Gamma_i + \frac{H}{2}\right)^2 = \left(\frac{G}{2}\right)^2 + \left(\frac{H}{2}\right)^2 - I, \tag{4.39}$$

$$\left(\Gamma_r + \frac{J}{2}\right)^2 + \left(\Gamma_i + \frac{K}{2}\right)^2 = \left(\frac{J}{2}\right)^2 + \left(\frac{K}{2}\right)^2 - L. \tag{4.40}$$

In (4.39) and (4.40), the parameters G, H, I, J, K, and L are defined below:

$$G = \frac{2g_L + p - u}{g_L + p}, \tag{4.41}$$

$$H = \frac{-(v + q)}{g_L + p}, \tag{4.42}$$

$$I = \frac{g_L - u}{g_L + p}, \tag{4.43}$$

$$J = \frac{2b_L + q - v}{b_L + q}, \tag{4.44}$$

$$K = \frac{p + u}{b_L + q}, \tag{4.45}$$

$$L = \frac{b_L - v}{b_L + q}. \tag{4.46}$$

It can be shown that (4.39) and (4.40) can be reduced to the conductance and susceptance circles of the Y Meta-Smith charts for CCITLs (see (4.15) and (4.16)), respectively, when $Z_0^+ = (Z_0^-)^*$ as well. Thus, the Y Meta-Smith charts for CCITLs are special cases of the Y Meta-Smith chart for BCITLs when $Z_0^+ = (Z_0^-)^*$ indeed.

   After deriving these four circle equations (see (4.31), (4.32), (4.39), and (4.40)), the Meta-Smith chart for BCITLs with NNCRs and NCRs can be constructed readily. In the next subsection, numerical results of the Meta-Smith chart for BCITLs will be illustrated.

4.4.3   NUMERICAL RESULTS

In this subsection, two examples of the Meta-Smith chart for BCITLs with NNCRs and NCRs are provided. First, consider a BCITL with NNCRs possessing $Z_0^+ = 149.3 - j7.9$ Ω and $Z_0^- = 146.9 - j39.6$ Ω, where Z_0^+ and Z_0^- are different with $(\text{Re}\{Z_0^\pm\} \geq 0)$. This BCITL can be implemented using a finite lossy periodic TL structure discussed in detail later in Section 5.6. Substituting the given $Z_0^\pm$ into (4.31) and (4.32), the resistance and reactance circles can be plotted as shown in Figures 4.11 and 4.12, respectively. In addition, Figure 4.13 is the Z Meta-Smith chart for BCITLs with NNCRs, which is the superposition of Figures 4.11 and 4.12. For resistance circles, the circles with $r_L > 0$ are inside the circle $r_L = 0$ while those with $r_L < 0$ are outside the circle $r_L = 0$. For reactance circles, the contours above $x_L = 0$ correspond to $x_L > 0$ and those below $x_L = 0$ correspond to $x_L < 0$. Note that the resistance and reactance circles intersect at the right angle due to the conformal mapping property.

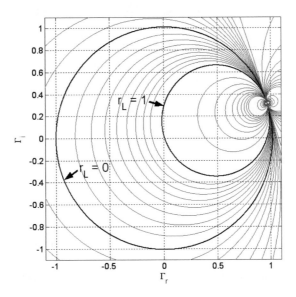

Figure 4.11: Plot of resistance circles for $Z_0^+ = 149.3 - j7.9\ \Omega$ and $Z_0^- = 146.9 - j39.6\ \Omega$.

Using (4.39) and (4.40), the Y Meta-Smith chart for BCITLs with NNCRs, which is the superposition of conductance and susceptance circles, can be constructed as illustrated in Figure 4.14. For conductance circles, the circles with $g_L > 0$ are inside the circle $g_L = 0$ while those with $g_L < 0$ are outside the circle $g_L = 0$. For susceptance circles, the contours above $b_L = 0$ correspond to $b_L < 0$, and those below $b_L = 0$ correspond to $b_L > 0$. Note that the conductance and susceptance circles intersect at the right angle as well.

Figure 4.15 shows the ZY Meta-Smith chart for BCITLs with NNCRs for $Z_0^+ = 149.3 - j7.9\ \Omega$ and $Z_0^- = 146.9 - j39.6\ \Omega$ by superimposing the Z Meta-Smith chart in Figure 4.13 and the Y Meta-Smith chart in Figure 4.14 together. From Figure 4.15, it is observed that the circle $x_L = 0$ and the circle $b_L = 0$ are the same. This can be verified by comparing (4.32) and (4.40) when $x_L = 0$ and $b_L = 0$, respectively. In addition, the circle $x_L = 0$ also joins the OC point and the SC point as shown in Figure 4.15.

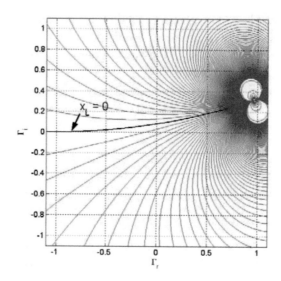

Figure 4.12: Plot of reactance circles for $Z_0^+ = 149.3 - j7.9\ \Omega$ and $Z_0^- = 146.9 - j39.6\ \Omega$.

Next, consider a BCITL with NCRs possessing $Z_0^+ = -35.825 - j12.480\ \Omega$ and $Z_0^- = -24.785 - j18.261\ \Omega$, where Z_0^+ and Z_0^- are different with $\text{Re}\{Z_0^\pm\} < 0$. This BCITL can be implemented using a finite lossy periodic TL structure [15]. Substituting the given $Z_0^\pm$ into (4.31) and (4.32), the resistance and reactance circles can be plotted together in the Γ plane resulting in the Z Meta-Smith chart for BCITLs with NCRs as shown in Figure 4.16. For resistance circles, the circles with $r_L < 0$ are inside the circle $r_L = 0$ while those with $r_L > 0$ are outside the circle $r_L = 0$. Note that this characteristic of resistance circles is opposite to that of the Z Meta-Smith chart for BCITLs with NNCRs as shown earlier in Figure 4.13. For reactance circles, the contours below $x_L = 0$ correspond to $x_L > 0$ and those above $x_L = 0$ correspond to $x_L < 0$. It is observed that this characteristic of reactance circles is opposite to that of the Z Meta-Smith chart for BCITLs with NNCRs as shown in Figure 4.13. Note that the resistance and reactance circles in Figure 4.16 intersect at the right angle.

Using (4.39) and (4.40), the Y Meta-Smith chart for BCITLs with NCRs, which is the superposition of conductance and susceptance circles, can be constructed as shown in Figure 4.17. For conductance circles, the circles with $g_L < 0$ are inside the circle $g_L = 0$ while those with $g_L > 0$ are outside the circle $g_L = 0$. For susceptance circles, the contours below $b_L = 0$ correspond to $b_L < 0$, and those above $b_L = 0$ correspond to $b_L > 0$. Note that these characteristics of conductance and susceptance circles are opposite to those of the Y Meta-Smith chart for BCITLs with NNCRs as shown earlier in Figure 4.14. In addition, it is observed that the conductance and susceptance circles in Figure 4.17 intersect at the right angle.

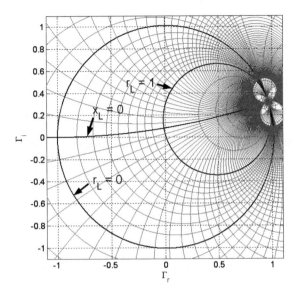

Figure 4.13: The Z Meta-Smith chart for BCITLs with NNCRs for $Z_0^+ = 149.3 - j7.9 \, \Omega$ and $Z_0^- = 146.9 - j39.6 \, \Omega$.

Figure 4.18 shows the ZY Meta-Smith chart for BCITLs with NCRs for $Z_0^+ = -35.825 - j12.480 \, \Omega$ and $Z_0^- = -24.785 - j18.261 \, \Omega$ by superimposing the Z Meta-Smith chart in Figure 4.16 and the Y Meta-Smith chart in Figure 4.17 together. From Figure 4.18, it is observed that the circle $x_L = 0$ and the circle $b_L = 0$ are identical. In addition, the circle $x_L = 0$ also joins the OC point and the SC point as shown in Figure 4.18.

4.5 CONCLUSIONS

In this chapter, the Meta-Smith charts for CCITLs with NNCRs and NCRs are developed, including the Meta-Smith chart for BCITLs. For the Meta-Smith chart for CCITLs with NNCRs, the region of interest is *inside or on* the unit circle in the Γ plane for passive load terminations. For the case of CCITLs with NCRs, the plotting range of the Meta-Smith chart is *always outside or on* the unit circle in the Γ plane for passive loads. It is found that the Meta-Smith charts for CCITLs with NNCRs and NCRs depend on the argument ϕ of $Z_0^\pm$ in a complicated fashion. In addition, the Smith chart is a special case for the Meta-Smith chart for CCITLs with NNCRs when $\phi = 0°$. Examples are also provided for these Meta-Smith charts. For the Meta-Smith chart for BCITLs, it is found that the magnitude of the voltage reflection coefficient at the load $|\Gamma|$ may be greater than unity for some passive load impedances Z_L. Thus, associated circles of the Meta-Smith chart for BCITLs are not only plotted inside the unit circle *but as well as outside* of the unit circle in the Γ plane. Two examples of the Meta-Smith chart for BCITLs with NNCRs and NCRs are provided.

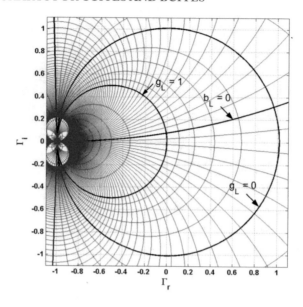

Figure 4.14: The Y Meta-Smith chart for BCITLs with NNCRs for $Z_0^+ = 149.3 - j7.9 \, \Omega$ and $Z_0^- = 146.9 - j39.6 \, \Omega$.

These Meta-Smith charts are very useful in the analysis and design of CCITLs and BCITLs since they can provide more physical insight of visualizing associated TL phenomenon and solving related problems effectively. In the next chapter, several applications of the Meta-Smith charts for CCITLs and BCITLs will be illustrated in detail.

BIBLIOGRAPHY

[1] D. Torrungrueng and C. Thimaporn, "A generalized *ZY* Smith chart for solvingnonreciprocal uniform transmission line problems," *Microwave and Optical Technology Letters*, vol. 40, no. 1, pp. 57–61, Jan. 2004. DOI: 10.1002/mop.11284 41, 42, 43, 44

[2] D. Torrungrueng, P.Y. Chou, and M. Krairiksh, "An extended ZY T-chart for conjugately characteristic-impedance transmission lines with active characteristic impedances," *Microwave and Optical Technology Letters*, vol. 49, no. 8, pp. 1961–1964, Aug. 2007. DOI: 10.1002/mop.22626 41, 48, 49

[3] D. Torrungrueng, P.Y. Chou, and M. Krairiksh, "A graphical tool for analysis and designof bi-characteristic-impedance transmission lines," *Microwave and Optical Technology Letters*, vol. 49, no. 10, pp. 2368–2372, Oct. 2007. DOI: 10.1002/mop.22801 53

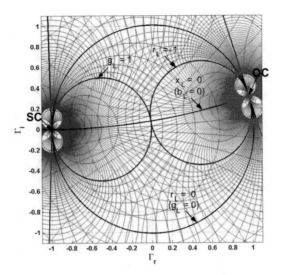

Figure 4.15: The ZY Meta-Smith chart for BCITLs with NNCRs for $Z_0^+ = 149.3 - j7.9 \ \Omega$ and $Z_0^- = 146.9 - j39.6 \ \Omega$.

[4] D. Torrungrueng, P.Y. Chou, and M. Krairiksh, "Erratum: A graphical tool for analysis and design of bi-characteristic-impedance transmission lines," *Microwave and Optical Technology Letters*, vol. 51, no. 4, pp. 1154, Apr. 2009. DOI: 10.1002/mop.24260 53

[5] D. Torrungrueng and C. Thimaporn, "Applications of the ZY T-chart for nonreciprocal stub tuners," *Microwave and Optical Technology Letters*, vol. 45, no. 3, pp. 259–262, May 2005. DOI: 10.1002/mop.20789

[6] D. Torrungrueng and C. Thimaporn, "Application of the T-chart for solving exponentially tapered lossless nonuniform transmission line problems," *Microwave and Optical Technology Letters*, vol. 45, no. 5, pp. 402–406, Jun. 2005. DOI: 10.1002/mop.20836

[7] D. Torrungrueng, C. Thimaporn, and N. Chamnandechakun, "An application of the T-Chart for solving problems associated with terminated finite lossless periodic structures," *Microwave and Optical Technology Letters*, vol. 47, no. 6, pp. 594–597, Dec. 2005. DOI: 10.1002/mop.21239 41

[8] S. Lamultree and D. Torrungrueng, "On the characteristics of conjugately characteristic-impedance transmission lines with active characteristic impedance," in *2006 Asia-Pacific Microwave Conference Proceedings*, Dec. 2006, vol. 1, pp. 225–228. DOI: 10.1109/APMC.2006.4429411 41

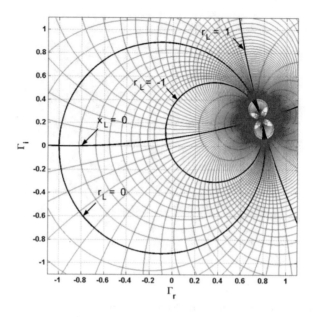

Figure 4.16: The Z Meta-Smith chart for BCITLs with NCRs for $Z_0^+ = -35.825 - j12.480$ Ω and $Z_0^- = -24.785 - j18.261$ Ω.

[9] K. Vudhivorn, D. Torrungrueng and C. Thimaporn, "Non-uniqueness of T-charts forsolving CCITL problems with passive characteristic impedances," *Progress inElectromagnetics Research*, vol. 7, pp. 193–205, 2009.
DOI: 10.2528/PIERM09052803 42, 46, 47, 48

[10] P. H. Smith, *Electronic Applications of the Smith Chart*. Georgia: Noble Publishing, 2000. 42, 45, 46

[11] D. M. Pozar, *Microwave Engineering*, 3rd ed. New Jersey: Wiley, 2005. 46

[12] E. Kreyszig, *Advanced Engineering Mathematics*, 8th ed. New Jersey: Wiley, 1998. 52

[13] K. Vudhivorn and D. Torrungrueng, "A modified extended ZY T-chart for conjugatelycharacteristic-impedance transmission lines with active characteristic impedances,"*Microwave and Optical Technology Letters*, vol. 51, no. 3, pp. 621–625, Mar. 2009.
DOI: 10.1002/mop.24124 52

[14] K. Vudhivorn, D. Torrungrueng and T. Angkaew, "Applications of the Modified Extended ZY T-Chart for Problems Associated with CCITLs with Active CharacteristicImpedances," in *Proceedings of the 2009 ECTI International Conference*, May 2009, vol. 2, pp. 822–825.
DOI: 10.1109/ECTICON.2009.5137172 52

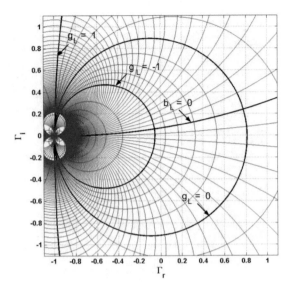

Figure 4.17: The Y Meta-Smith chart for BCITLs with NCRs for $Z_0^+ = -35.825 - j12.480\ \Omega$ and $Z_0^- = -24.785 - j18.261\ \Omega$.

[15] S. Lamultree, D. Torrungrueng and C. Phongcharoenpanich, "Measurement of BCITL parameters," in *Proceedings of the 2008 ECTI International Conference*, May 2008, vol. 1, pp. 257–260. DOI: 10.1109/ECTICON.2008.4600421 58

4.6   PROBLEMS

4.1. Using (4.9), show that the equations for resistance and reactance circles for CCITLs are given as in (4.12) and (4.13), respectively.

4.2. Show that the equations for conductance and susceptance circles for CCITLs are given as in (4.15) and (4.16), respectively.

4.3. Rigorously show that the relationship between the argument ϕ and the angle θ (see Figure 4.2 as an example) for CCITLs with NNCRs is given as in (4.17).

4.4. Plot the ZY Meta-Smith chart for CCITLs with NNCRs in the Γ plane for $\phi = 0°, \pm30°$ and $\pm60°$ (see Figure 4.5 as an example).

4.5. Using $\tilde{Z}_0$ defined in (4.18), show that the equations for resistance, reactance, conductance and susceptance circles for CCITLs with NNCRs are given as in (4.19) to (4.22), respectively.

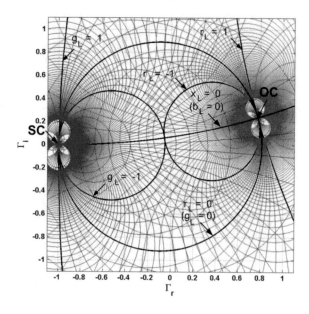

Figure 4.18: The ZY Meta-Smith chart for BCITLs with NCRs for $Z_0^+ = -35.825 - j12.480$ Ω and $Z_0^- = -24.785 - j18.261$ Ω.

4.6. Using (4.19) to (4.22), plot the ZY Meta-Smith chart for CCITLs with NNCRs in the Γ plane for $\phi = 0°, \pm 30°$ and $\pm 60°$. Compare these results with those of Problem 4.4.

4.7. Rigorously show that the relationship between the argument ϕ and the angle θ (see Figure 4.6 as an example) for CCITLs with NCRs is given as in (4.25).

4.8. Plot the ZY Meta-Smith chart for CCITLs with NCRs in the Γ plane for $\phi = 120°$, $\pm 135°$ and $\pm 150°$ (see Figure 4.9 as an example).

4.9. Using (4.28), show that the equations for resistance and reactance circles for BCITLs are given as in (4.31) and (4.32), respectively.

4.10. Show that (4.31) and (4.32) can be reduced to the equations of the resistance and reactance circles of the Z Meta-Smith charts for CCITLs (see (4.12) and (4.13)) respectively, when $Z_0^+ = \left(Z_0^-\right)^*$.

4.11. Show that the equations for conductance and susceptance circles for BCITLs are given as in (4.39) and (4.40), respectively.

4.12. Verify the plot of the ZY Meta-Smith chart for BCITLs with NNCRs in Figure 4.15 for $Z_0^+ = 149.3 - j7.9$ Ω and $Z_0^- = 146.9 - j39.6$ Ω.

4.13. Verify the plot of the ZY Meta-Smith chart for BCITLs with NCRs in Figure 4.18 for $Z_0^+ = -35.825 - j12.480 \ \Omega$ and $Z_0^- = -24.785 - j18.261 \ \Omega$.

CHAPTER 5

Applications of Meta-Smith Charts

5.1 INTRODUCTION

In Chapter 4, the Meta-Smith charts for CCITLs and BCITLs with NNCRs and NCRs are developed. These charts usually provide more physical insight of visualizing phenomenon associated CCITLs and BCITLs and solving related problems effectively as can be seen in several examples, involving impedance matching and impedance calculation associated with CCITLs and BCITLs, discussed in this chapter. It will be shown later in this chapter that graphical solutions based on computerized Meta-Smith charts are fast, intuitive and accurate, compared to analytical solutions.

This chapter is organized as follows. Nonreciprocal stub tuners are designed using the Meta-Smith chart for CCITLs with NNCRs for both single-stub and double-stub matching networks in Section 5.2. Sections 5.3 and 5.4 illustrate the usage of the Meta-Smith chart for CCITLs with NNCRs in solving problems associated with exponentially tapered lossless nonuniform transmission lines and terminated finite reciprocal lossless periodic TL structures with NNCRs, respectively. A CCITL with NCRs is analyzed using the Meta-Smith chart for CCITLs with NCRs in Section 5.5. Section 5.6 presents the usage of the Meta-Smith chart for BCITLs with NNCRs in solving a problem associated with reciprocal lossy BCITLs with NNCRs. Finally, conclusions are provided in Section 5.7.

5.2 NONRECIPROCAL STUB TUNERS

Impedance matching is a part of design process for a microwave system to achieve the maximum power delivered to the load. Stub-tuning is one of conventional matching techniques, which is usually implemented using single-stub and double-stub tuners. A stub is an open- or short-circuited TL connected either in parallel or in series with the transmission feed line at a certain distance from the load [1]. In this section, the Meta-Smith chart for CCITLs with NNCRs is applied for solving problems associated with nonreciprocal stub tuners composed of nonreciprocal lossless uniform transmission lines, where these TLs are CCITLs with NNCRs. Both nonreciprocal single-stub and double-stub matching networks are studied graphically using the Meta-Smith chart for CCITLs with NNCRs. The stubs can be connected in series or in parallel with the main nonreciprocal lossless feed line. It will show that basic procedures of using the Meta-Smith chart in solving problems of nonreciprocal stub tuners are similar to those of applying the Smith chart for reciprocal ones discussed in [1]. It should be pointed out that nonreciprocal stub tuners can be designed using analytical and graphical solutions as well [2]– [3]. Although analytical solutions are more accurate and useful for computer analysis and design, they are complicated [2]. On the other hand, graphical solutions based on the Meta-Smith chart are found to be fast, intuitive, and sufficiently accurate in practice [3], which will be illustrated subsequently.

In this section, it is assumed for simplicity that all TLs of interest are *nonreciprocal* and *lossless*, possessing the same characteristic impedances $Z_0^\pm = |Z_0^\pm| e^{\mp j\phi}$ with NNCRs (Re$\{Z_0^\pm\} \geq 0$), and the same propagation constants $\beta^\pm$. First, graphical solutions based on the Meta-Smith chart for CCITLs with NNCRs for nonreciprocal single-stub series tuners are discussed in Section 5.2.1. Next, the Meta-Smith chart for CCITLs with NNCRs is applied for nonreciprocal double-stub shunt tuners as illustrated in Section 5.2.2. It should be pointed out that the Meta-Smith chart can be applied for nonreciprocal stub tuners when each nonreciprocal TL of interest is different as well.

5.2.1   NONRECIPROCAL SINGLE-STUB SERIES TUNERS

In the single-stub series tuning, the design parameters are the distance d from the load to the stub and the length ℓ of the OC or SC stub as shown in Figure 5.1. The single-stub series tuning circuit can be designed using the Z Meta-Smith chart for CCITLs with NNCRs. In order to match a load

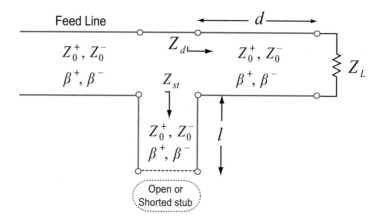

Figure 5.1: A nonreciprocal single-stub series tuner.

with the input nonreciprocal feed line, it is required to make the voltage reflection coefficient looking into the matching network equal to zero. Using (2.7) for the desired voltage reflection coefficient, the above requirement can be achieved by choosing the distance d and the length ℓ such that

$$Z_0^+ = Z_d + Z_{st}, \tag{5.1}$$

where Z_d is the input impedance looking into the nonreciprocal TL terminated in a passive load Z_L, and Z_{st} is the input impedance looking into the nonreciprocal stub as shown in Figure 5.1.

Using (2.7) and (2.8), it can be shown that Z_{st} is *purely imaginary* due to OC or SC termination of the stub. The normalized version of (5.1) can be written as

$$z_0^+ = z_d + z_{st}, \tag{5.2}$$

where z_0^+, z_d, and z_{st} are the normalized impedances (see (4.4) and (4.5)) of Z_0^+, Z_d, and Z_{st}, respectively. The Meta-Smith chart solutions can be illustrated best by an example as shown below.

Consider an example of a single-stub series tuning network implemented using an *OC* stub when $Z_L = 50\ \Omega$, $|Z_0^+| = 50\ \Omega$ and $\phi = 45°$. Let us define $\tilde{\beta}$ be the *arithmetic* mean of $\beta^{\pm}$; i.e.,

$$\tilde{\beta} \equiv \frac{1}{2}\left(\beta^+ + \beta^-\right) \equiv \frac{2\pi}{\tilde{\lambda}}, \tag{5.3}$$

where $\tilde{\lambda}$ is the *effective* wavelength of waves propagating along NRLUTLs. From the given data, the normalized load impedance z_L is equal to 1. For this case, the Z Meta-Smith chart for CCITLs with NNCRs is employed with $\phi = 45°$ as shown in Figure 5.2.

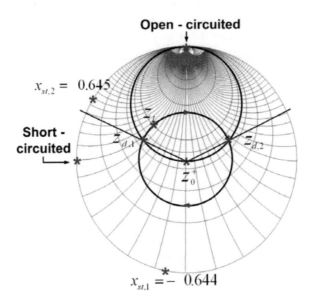

Figure 5.2: The Z Meta-Smith chart solutions for a single-stub series tuner with $\phi = 45°$.

The first step is to draw a circle, centered at the point $\Gamma = 0$, passing through the point z_L on the Z Meta-Smith chart for CCITLs with NNCRs. It should be pointed out that the point $\Gamma = 0$ corresponds to the point z_0^+, which can be seen from (4.6) and (4.7). Note that the circle intersects the resistance circle, passing through the point $\Gamma = 0$, in the Γ plane at two intersection points denoted as $z_{d,1}$ and $z_{d,2}$ as shown in Figure 5.2. Let us consider the first solution corresponding to $z_{d,1}$ first, where $z_{d,1}$ can be read from the computerized Z Meta-Smith chart as $z_{d,1} = 0.708 - j0.063$. From Figure 5.2, the shortest distance d_1 from the load to the nonreciprocal stub can be determined as $d_1 = [0.5 - (0.063 - 0.034)]\tilde{\lambda} = 0.471\ \tilde{\lambda}$. In order to determine the required stub length ℓ, one equates the imaginary part of (5.2); i.e.,

$$x_{st,1} = Im\{z_0^+\} - x_{d,1}, \tag{5.4}$$

where $x_{st,1}$ and $x_{d,1}$ are the normalized reactances of $z_{st,1}$ and $z_{d,1}$ respectively, and $z_{st,1}$ is the normalized impedance z_{st} of the first solution. Note that the symbol Im{} in (5.4) denotes the imaginary part of its argument. From (5.4) with $z_{d,1} = 0.708 - j0.063$, the first solution requires a nonreciprocal stub with the normalized reactance of $x_{st,1} = -0.644$. The length ℓ_1 of an OC stub that provides $x_{st,1}$ can be found using the computerized Z Meta-Smith chart by starting at the OC point, and moving along the outer edge of the Z Meta-Smith chart, corresponding to the resistance circle of $r_L = 0$ (see (4.12)), *toward the generator* (in the *clockwise* direction) to the point $x_{st,1}$ as shown in Figure 5.2. This provides a stub length of $\ell_1 = 0.264\ \tilde{\lambda}$. Similarly, the distance d_2 from the load and the required OC stub length ℓ_2 of the second solution are found to be: $d_2 = 0.153\ \tilde{\lambda}$ and $\ell_2 = 0.423\ \tilde{\lambda}$. It is interesting to point out that these results are sufficiently accurate compared with analytical solutions [2]. This completes the nonreciprocal single-stub series tuner design. Note that the nonreciprocal single-stub *shunt* tuner can be employed as well by connecting the nonreciprocal stub in parallel instead. Its procedures of using the Meta-Smith chart are similar to those of the series one. In Section 5.2.2, the design of nonreciprocal double-stub shunt tuners is discussed in detail.

5.2.2   NONRECIPROCAL DOUBLE-STUB SHUNT TUNERS

In nonreciprocal double-stub shunt tuning, the lengths of the two nonreciprocal OC or SC stubs are required, where the distance d between them is fixed as shown in Figure 5.3. Note that the load admittance Y_L is located at the position of the first nonreciprocal stub. In Figure 5.3, B_1 and B_2 are the input susceptances of the first and second nonreciprocal stubs, respectively. The Meta-Smith chart solutions are illustrated by the following example.

Consider a nonreciprocal double-stub shunt tuning network implemented using *SC* nonreciprocal stubs when $Y_L = 10 + j50\,S$, $|Y_0^\pm| = 50\,S$, $\phi = 45°$, and $d = \tilde{\lambda}/8$, where characteristic admittances of all nonreciprocal transmission lines with NNCRs are defined as $Y_0^\pm = 1/Z_0^\pm = |Y_0^\pm|e^{\pm j\phi}$ with $|Y_0^\pm| = 1/|Z_0^\pm|$. From the given data, the normalized load admittance $y_L = Y_L/|Y_0^\pm|$ is equal to 0.2 $+ j$. As in the case of the nonreciprocal single-stub tuners, two common solutions are possible. After the normalized load admittance y_L is plotted on the Y Meta-Smith chart with $\phi = 45°$ as shown in Figure 5.4, the conductance circle passing through the point $\Gamma = 0$ is rotated by moving every point on this circle $\tilde{\lambda}/8$ *toward the load* (in the *counterclockwise* direction). Then, the normalized input susceptance b_1 of the first solution of the first nonreciprocal stub can be obtained by starting at the point y_L and moving along the conductance circle, passing through y_L, to the intersection point y_1 between the conductance circle and the rotated circle as shown in Figure 5.4. Using the comput-

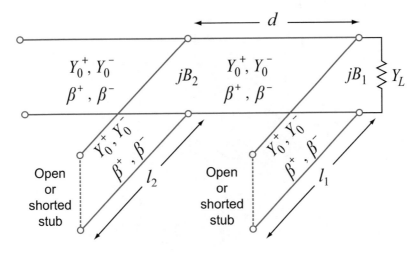

Figure 5.3: A nonreciprocal double-stub shunt tuner.

erized Y Meta-Smith chart, it is found that $b_1 = 0.908$. Now one transforms y_1 through the $\tilde{\lambda}/8$ section of nonreciprocal transmission line by rotating along a constant radius circle (passing through y_1) $\tilde{\lambda}/8$ *toward the generator* (in the *clockwise* direction). This brings the first solution y_1 to the point $y_2 = 0.706 - j1.733$. Then, the normalized input susceptance b_2 of the second nonreciprocal stub of the first solution is obtained by starting at the point y_2 and moving along the conductance circle (passing through the point $\Gamma = 0$) to the point y_0^+ as shown in Figure 5.4. It should be pointed out that the point $\Gamma = 0$ corresponds to the point y_0^+. Using the computerized Y Meta-Smith chart, it is found that $b_2 = 2.441$.

Once b_1 and b_2 are known, the lengths of the first and second SC nonreciprocal stubs for the first solution are found from the computerized Y Meta-Smith chart to be $\ell_1 = 0.294\ \tilde{\lambda}$ and $\ell_2 = 0.438\ \tilde{\lambda}$, respectively. Similarly, the second solution can be determined as $\ell_1 = 0.116\ \tilde{\lambda}$ and $\ell_2 = 0.0615\ \tilde{\lambda}$. This completes the nonreciprocal double-stub shunt tuning design. Note that the nonreciprocal double-stub series tuning circuit can be used as well by connecting the nonreciprocal stubs in series instead.

For a double-stub tuner, there is a specific region, where the load cannot be matched, called *a forbidden region* [1]. For convenience, only shunt stubs are considered. From the above discussion, it is obvious that if the normalized load admittance y_L were inside the *shaded* region, called the forbidden region, as shown in Figure 5.4, no value of shunt double-stub susceptances could bring the point y_L to intersect the rotated conductance circle. It should be pointed out that the forbidden region can be identified obviously on the Meta-Smith chart. This is an advantage of using the Meta-Smith chart in this problem. From analytical solutions of the nonreciprocal double-stub shunt tuning, it can be shown rigorously that the equation of the conductance circle governing the forbidden region can be

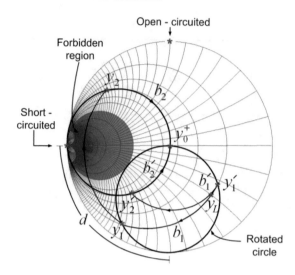

Figure 5.4: A Y Meta-Smith chart diagram for the operation of a nonreciprocal double-stub shunt tuner with $\phi = 45°$.

expressed compactly as

$$\left(\Gamma_r + \frac{1}{1 + \sin^2 \tilde{\beta}d}\right)^2 + \Gamma_i^2 = \left(\frac{\sin^2 \tilde{\beta}d}{1 + \sin^2 \tilde{\beta}d}\right)^2. \tag{5.5}$$

Comparing (5.5) to (4.15), it is found that the conductance circle equation in (5.5) corresponds to the normalized conductance of $g = g_{\max}$, where

$$g_{\max} = \frac{\cos \phi}{\sin^2 \tilde{\beta}d}. \tag{5.6}$$

From (5.5), it is interesting to point out that the forbidden region for the nonreciprocal double-stub shunt tuners depends only on the phase $\tilde{\beta}d$. In this example, the normalized load conductance g_L is equal to 0.2, which is less than $g_{\max} = \sqrt{2}$ computed via (5.6). Thus, the nonreciprocal double-stub shunt tuners can be applied to match the given load to the nonreciprocal feed line as shown earlier.

In conclusion, nonreciprocal stub tuners can be designed using analytical and Meta-Smith chart solutions. Although analytical solutions are more accurate and useful for computer analysis, they are quite complicated and difficult to obtain for nonreciprocal stub tuners. In contrast, solutions based on the computerized Meta-Smith chart are fast, intuitive and accurate. It is found that basic procedures and concepts for designing nonreciprocal stub tuners are similar to those of reciprocal ones. In the next section, the Meta-Smith chart for CCITLs with NNCRs is applied to solve problems associated with ETLNUTLs.

5.3 TERMINATED EXPONENTIALLY TAPERED LOSSLESS NONUNIFORM TRANSMISSION LINES

Due to complicated equations associated with terminated ETLNUTLs as shown in Figure 1.4, a graphical solution assisting in analysis and design is desirable. In the literature [4], a generalized Smith chart, which is one of the graphical solutions, was developed and employed to study ETLNUTLs. However, this chart is not convenient in solving ETLNUTL problems due to its complicated graphical representation.

Due to the fact that practical ETLNUTLs are reciprocal CCITLs with NNCRs when TL parameters and frequency are chosen appropriately (see Section 2.4), the Meta-Smith chart for CCITLs with NNCRs can be employed to solve ETLNUTL problems effectively. It should be pointed out that this Meta-Smith chart is totally different from another graphical tool proposed in [4], and it is more convenient and intuitive in solving these problems. Note that this Meta-Smith chart depends on the argument ϕ of the effective characteristic impedances $Z_0^{\pm}$ of ETLNUTLs defined earlier in (1.65) and (1.66), where $Z_0^{\pm}$ are dependent on parameters of ETLNUTLs and frequency. It will be illustrated that the procedures of using this Meta-Smith chart are similar to those of using the standard ZY Smith chart in solving problems associated with RLUTLs. As pointed out earlier in Section 4.2, this Meta-Smith chart can be reduced to the standard ZY Smith chart when $\phi = 0°$; i.e., ETLNUTLs become RLUTLs in this case.

To apply the Meta-Smith chart for CCITLs with NNCRs, the input impedance equation of terminated ETLNUTLs given in (1.72) must be rearranged into the following convenient form [5]:

$$\tilde{Z}_{in} \equiv Z_{in}e^{q\ell} = Z_0^{+}Z_0^{-}\left[\frac{1 + \Gamma e^{-j2\beta\ell}}{Z_0^{-} - Z_0^{+}\Gamma e^{-j2\beta\ell}}\right], \tag{5.7}$$

where $\tilde{Z}_{in}$ is the *modified* input impedance of the ETLNUTL terminated in a passive real load impedance Z_L. Notice that the exponential term $e^{-q\ell}$ in (1.72) is moved to the left-hand side as shown in (5.7). It is found that (5.7) is in the same form as the equation of the input impedance of CCITLs terminated in a load impedance Z_L (see (2.26)). Thus, the Meta-Smith chart for CCITLs with NNCRs can be applied to solve ETLNUTL problems indeed by working with the modified input impedance $\tilde{Z}_{in}$ rather than the input impedance Z_{in}.

5.3.1 PROCEDURES OF USING THE META-SMITH CHART

To apply the Meta-Smith chart for CCITLs with NNCRs, the *normalized* impedance z of ETLNUTLs is employed, and it is defined as in (4.4) and (4.5). Thus, the normalized modified input impedance $\tilde{z}_{in}$ of $\tilde{Z}_{in}$ defined in (5.7) can be expressed as

$$\tilde{z}_{in} = \frac{1 + \Gamma e^{-j2\beta\ell}}{z_0^{-} - z_0^{+}\Gamma e^{-j2\beta\ell}}, \tag{5.8}$$

where $z_0^{\pm}$ are the normalized effective characteristic impedances defined in (4.7). Once $\tilde{z}_{in}$ is known, the *modified* input impedance $\tilde{Z}_{in}$ can be obtained by using (4.4) and (4.5), and the input impedance

Z_{in} can be readily determined from (5.7) as

$$Z_{in} = \tilde{Z}_{in}e^{-q\ell}. \tag{5.9}$$

The procedures of determining $\tilde{z}_{in}$ by using the Meta-Smith chart for CCITLs with NNCRs are described below.

Consider the normalized load impedance z_L defined as in (4.8). One can see that (5.8) for $\tilde{z}_{in}$ is in the same form as (4.8), except for the extra phase factor $e^{-j2\beta l}$ multiplying Γ for both numerator and denominator in (5.8). For convenience in discussion, define the voltage reflection coefficient Γ at the load in the polar form as $|\Gamma|e^{j\psi}$, where $|\Gamma|$ and ψ are the absolute value and the argument of Γ, respectively. The first step in applying the Meta-Smith chart for CCITLs with NNCRs is to plot the reflection coefficient at the load $|\Gamma|e^{j\psi}$ on the chart, which corresponds to the normalized load impedance point z_L. Then, the normalized modified input impedance $\tilde{z}_{in}$ can be determined by simply rotating the z_L point *clockwise* (toward the generator) an amount of $2\beta\ell$ (i.e., subtracting $2\beta\ell$ from ψ) around the origin of the Meta-Smith chart. One can see that the above procedures of applying the Meta-Smith chart for CCITLs with NNCRs in solving ETLNUTL problems are similar to those of using the standard Smith chart in solving problems associated with RLUTLs. These procedures of using the Meta-Smith chart for CCITLs with NNCRs can be seen more clearly by considering a numerical example given in the next subsection.

5.3.2   NUMERICAL EXAMPLE

In this subsection, an example will be provided to show that solutions obtained from the Meta-Smith chart for CCITLs with NNCRs are valid by comparing with results obtained from associated formulas. Consider the case when $C_0 = 30$ pF/m, $q = 2$ Np/m, $f = 1$ GHz, $L_0 = 1.688$ nH/m, $\beta\ell = 45°$ and $Z_L = 7.5$ Ω [5]. Using the given data and (1.62), (1.65), and (1.66), β and $Z_0^{\pm}$ can be computed as $\beta = 1$ rad/m and $Z_0^{\pm} = 7.5\,e^{\mp j45°}$ Ω. Note that $|Z_0^{\pm}|$ and ϕ are equal to 7.5 Ω and 45° in this case, respectively. Using (4.4) and (4.5), the normalized load impedance z_L is determined as $z_L = r_L + jx_L = 1$. In this case, $r_L = 1$ and $x_L = 0$ correspond to a real load, which can be plotted as the z_L point on the Meta-Smith chart for CCITLs with NNCRs with $\phi = 45°$ as shown in Figure 5.5. Next, a circle passing through the z_L point is drawn on the Meta-Smith chart. Then, the normalized modified input impedance $\tilde{z}_{in}$ can be found by rotating the z_L point *clockwise* along the circle by an amount of $2\beta\ell = 90°$. Note that the $\tilde{z}_{in}$ point can be read from the computerized Meta-Smith chart for CCITLs with NNCRs as $0.98 - j1.41$. Once $\tilde{z}_{in}$ is known, the modified input impedance $\tilde{Z}_{in}$ can be calculated as $\tilde{Z}_{in} = \tilde{z}_{in}|Z_0^{\pm}| = 7.35 - j10.57$ Ω, and the input impedance Z_{in} of the terminated ETLNUTL can be determined by using (5.9) as $Z_{in} = 1.527 - j2.197$ Ω. Alternatively, the input impedance Z_{in} can be determined analytically by using (1.72) as $Z_{in} = 1.559 - j2.204$ Ω. Comparing these results, it is found that both are in very good agreement. Thus, this example illustrates that the computerized Meta-Smith chart for CCITLs with NNCRs provides accurate results indeed, and it can be employed readily and intuitively in analysis and design of ETLNUTLs.

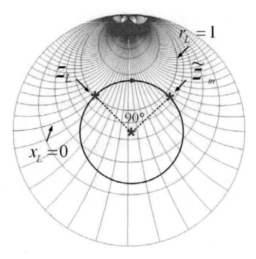

Figure 5.5: Illustration of the procedures of applying the Meta-Smith chart to solve an ETLNUTL problem with $\phi = 45°$.

5.4 TERMINATED FINITE RECIPROCAL LOSSLESS PERIODIC TRANSMISSION LINE STRUCTURES WITH NONNEGATIVE CHARACTERISTIC RESISTANCES

As discussed earlier in Section 2.5, periodic TL structures have several useful applications in microwave technology. In addition, lossless periodic TL structures are probably the most useful example of CCITLs in practice. Due to complicated equations involved with terminated finite reciprocal lossless periodic TL structures, a graphical solution assisting in analysis and design of these structures is desirable. In this section, the Meta-Smith chart for CCITLs with NNCRs is applied to solve problems associated with these structures operating in passbands readily and effectively, especially for *unsymmetrical* unit cells associated with periodic TL structures. The procedures of applying the Meta-Smith chart for these periodic TL structures are provided, including an example. Only the periodic TL structures with NNCRs are considered in this section.

Consider the terminated finite reciprocal lossless periodic TL structure of M unit cells in Figure 2.3. Using (2.28), (4.4), and (4.5), the normalized input impedance $z_{in,M}$ at the input terminal of the terminated periodic TL structure can be expressed as

$$z_{in,M} = \frac{1 + \Gamma e^{-j2M\beta d}}{z_0^- - z_0^+ \Gamma e^{-j2M\beta d}}, \tag{5.10}$$

where $z_0^\pm$ are the normalized effective characteristic impedances of $Z_0^\pm$. The normalized load impedance z_L can be obtained from (5.10) by setting $d = 0$ resulting in (4.8). Note that (4.8) is in the same form as (5.10), except for the extra phase factor $e^{-j2M\beta d}$ multiplying the Γ term. For convenience in discussion, let us define the voltage reflection coefficient Γ at the load as $|\Gamma|e^{j\psi}$. Thus, if one has plotted the reflection coefficient $|\Gamma|e^{j\psi}$ at the load, corresponding to the normalized load impedance point z_L on the Meta-Smith chart, the normalized input impedance $z_{in,M}$ can be determined by simply rotating the z_L point *clockwise* an amount of $2M\beta d$ (subtracting $2M\beta d$ from ψ) around the origin of the Meta-Smith chart for CCITLs with NNCRs. Now, it is ready to illustrate the use of the Meta-Smith chart through a specific example.

Let us consider the unit cell as illustrated in Figure 2.2 with the following parameters: $d = 2$ cm, $k = 120.323$ rad/m, $Z_1 = 13.928$ Ω, $C_1 = 1$ pF, $L = 2.7$ nH, $C = 5$ pF, $L_1 = 11$ nH and $f = 3.458$ GHz [6]. In this example, three unit cells ($M = 3$) terminated in $Z_L = 1.153 + j1.153$ Ω are employed for illustration. Using the cascading property of the *ABCD* matrix (see Appendix A), the total *ABCD* matrix for this case is found to be

$$\begin{bmatrix} A & B \\ C & D \end{bmatrix} = \begin{bmatrix} -0.749 & j1.857 \\ j0.349 & -0.469 \end{bmatrix}. \tag{5.11}$$

Using (2.20), (2.21), and (2.27), $Z_0^\pm$ and β are found to be $Z_0^\pm = 2.306\ e^{\pm j10°}$ Ω and $\beta = 37.09$ rad/m. Note that the argument ϕ of Z_0^- in this case is equal to $-10°$, and the finite periodic TL structure is operated in a passband since β is real. In addition, the effective propagation constant β of the periodic TL structure is significantly different from the propagation constant k of the *unloaded* transmission line in Figure 2.2.

Using (4.4) and (4.5), the normalized load impedance z_L for this example is found to be $z_L = r_L + jx_L = 0.5 + j0.5$. In this case, it is found that $r_L = 0.5$ and $x_L = 0.5$, which can be plotted as the z_L point on the Meta-Smith chart with $\phi = -10°$ as shown in Figure 5.6. Next, a circle passing through the z_L point is drawn on the Meta-Smith chart. On the chart, the normalized input impedance $z_{in,M}$ can be found by rotating the z_L point *clockwise* an amount of $2M\beta d = 255°$ around the origin of the chart (moving *toward the generator*) as shown in Figure 5.6. Thus, the $z_{in,M}$ point can be read as $0.54 - j0.24$, and the input impedance $Z_{in,M}$ can be calculated as

$$Z_{in,M} = z_{in,M}|Z_0^\pm| = 1.25 - j0.56\ \Omega. \tag{5.12}$$

Note that the analytical result obtained from (2.24) is equal to $Z_{in,M} = 1.2501 - j0.5604$ Ω, which is in very good agreement with the one obtained from the computerized Meta-Smith chart in (5.12). Therefore, the computerized Meta-Smith chart for CCITLs with NNCRs provides accurate results, and it can be used in analysis and design of terminated finite reciprocal lossless periodic TL structures indeed.

It should be pointed out that the Smith chart can be used to determine the input impedance $Z_{in,M}$ of the terminated finite periodic TL structure in this example as well. However, using the Smith chart for calculation is *not convenient* compared to using the Meta-Smith chart for CCITLs with NNCRs, especially when each unit cell is *complex* and the number of unit cells is *large*. In addition, the Smith chart does not provide physical insight in solving this problem compared to the

Meta-Smith chart since the information on periodicity of the TL structure is not employed in the Smith chart calculation at all. For this particular example, it is found that both the Meta-Smith chart for CCITLs with NNCRs and the Smith chart yield almost identical results as expected. In the next section, the Meta-Smith chart for CCITLs with NCRs is employed to solve problems associated with reciprocal CCITLs with NCRs.

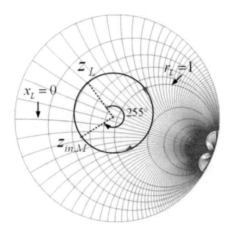

Figure 5.6: The Meta-Smith chart solution for the terminated finite periodic TL structure with NNCRs when $\phi = -10°$.

5.5 CCITLS WITH NEGATIVE CHARACTERISTIC RESISTANCES

Consider a reciprocal CCITL with NCRs of length $\lambda/8$ with $\phi = -150°$, terminated in a normalized load impedance of $z_L = 0.8 + j1.2$ as shown in Figure 2.4, where λ is the wavelength of waves propagating along the CCITL. Note that this reciprocal CCITL can be implemented using appropriate finite reciprocal lossless periodic TL structures [7]. Using (2.26), (4.4), and (4.5), the normalized input impedance of the terminated CCITL z_{in} can be expressed as

$$z_{in} = \frac{1 + \Gamma e^{-j2\beta\ell}}{z_0^- - z_0^+ \Gamma e^{-j2\beta\ell}}, \tag{5.13}$$

where $z_0^{\pm} = e^{\mp j\phi}$ are the normalized characteristic impedances of the CCITL. The procedures of using the Z Meta-Smith chart for a CCITL with NCRs for impedance calculation are similar to those of using the Z Meta-Smith chart for CCITLs with NNCRs given earlier in Section 5.4 due to the fact that the formulas for computing the normalized input impedance for both cases are in the same form.

Figure 5.7 illustrates the impedance calculation on the Z Meta-Smith chart for the CCITL with NCRs of the above example. Note that the operating region is *outside or on* the unit circle for *passive* load terminations as shown in Figure 5.7, and the magnitude of the reflection coefficient at the load $|\Gamma|$ in this case is found to be 2.57 (see (4.6)), which is greater than unity as expected. Starting at $z_L = 0.8 + j1.2$, move along the circle passing through z_L in the *clockwise* direction (moving *toward the generator*) for $\lambda/8(2\beta l = 90°)$ finally arrive at $z_{in} = 0.4 + j0.6$, which is very close to the result obtained from the formula in (5.13); i.e., $z_{in} = 0.388 + j0.606$.

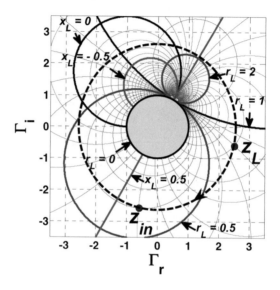

Figure 5.7: Impedance calculation on the Z Meta-Smith chart for the CCITL with NCRs of length $\lambda/8$ terminated in $z_L = 0.8 + j1.2$ when $\phi = -150°$.

5.6 RECIPROCAL LOSSY BCITLS

In this section, a finite reciprocal lossy periodic TL structure with NNCRs is employed as an example of reciprocal lossy BCITLs with NNCRs. Consider a reciprocal *lossy* unit cell of a finite lossy periodic TL structure terminated in a passive load impedance Z_L as illustrated in Figure 3.4 with $d = 2$ cm, the characteristic impedance $Z_1 = 150$ Ω, the wavelength along the reciprocal lossless uniform TL $\lambda = 7.5$ cm, a series resistance $R_S = 75$ Ω, a shunt resistance $R_{SH} = 200$ Ω and $Z_L = 50$ Ω [8], [9]. Only one unit cell ($M = 1$) is considered in this example. Note that the propagation constant k of the reciprocal lossless uniform TL is equal to $2\pi/\lambda$. In addition, the length (ℓ) of the equivalent BCITL is equal to d in this case. The total *ABCD* matrix of the unit cell for this case is readily found as

$$\begin{bmatrix} A & B \\ C & D \end{bmatrix} = \begin{bmatrix} 0.063 + j0.622 & -28.550 + j177.149 \\ j0.008 & -0.312 + j0.622 \end{bmatrix}. \tag{5.14}$$

Once the total $ABCD$ matrix is known, the characteristic impedances $Z_0^{\pm}$ and the complex propagation constant γ of the equivalent BCITL can be determined by using associated formulas given in Section 3.3. Using (3.10), (3.11), and (3.15), it is numerically found that $Z_0^+ = 149.3 - j7.9 \ \Omega$, $Z_0^- = 146.9 - j39.6 \ \Omega$ and $\gamma = \alpha + j\beta = 29.5 + j83.8$ 1/m. Note that Z_0^+ and Z_0^- are *not* complex conjugate of one another as expected since the network is *lossy*. Using (3.5), it is found that the magnitude of the reflection coefficient at the load $|\Gamma|$ is equal to 0.505.

Using (4.27), the normalized load impedance z_L in this example is found to be $z_L = r_L + jx_L = 0.330 - j0.035$. In this case, it is found that $r_L = 0.330$ and $x_L = -0.035$, which can be plotted as the z_L point on the Z Meta-Smith chart for BCITLs with NNCRs as shown in Figure 5.8. Due to the loss of the BCITL, an exponentially attenuating spiral starting from the z_L point is drawn on the Z Meta-Smith chart for BCITLs with NNCRs. On the chart, the normalized input impedance z_{in} can be found by rotating the z_L point *clockwise* an amount of $2\beta l = 192°$ along the spiral path due to the lossy nature of the BCITL (see (3.6) and (3.7) with $\gamma^+ = \gamma^- = \gamma$ for *reciprocal* lossy BCITLs). Thus, the z_{in} point can be read from the computerized Z Meta-Smith chart as $1.27 - j0.34$, and the input impedance Z_{in} can be computed from z_{in} as

$$Z_{in} = z_{in}\sqrt{Z_0^+ Z_0^-} = 195.93 - j30.91 \ \Omega. \tag{5.15}$$

Alternatively, using the standard cascading approach for the TL calculation in Figure 3.4, it is found that $Z_{in} = 196.09 - j31.35 \ \Omega$, which is in very good agreement with that obtained from the computerized Z Meta-Smith chart for BCITLs with NNCRs in (5.15). Therefore, the chart can provide sufficiently accurate results indeed. Note that the usage of the Z Meta-Smith chart for solving BCITLs with NNCRs as shown above is similar to that of the standard Smith chart for solving RLSUTLs.

5.7   CONCLUSIONS

In this chapter, the Meta-Smith charts for CCITLs and BCITLs developed in Chapter 4 are applied to solve TL problems involving impedance matching and impedance calculation. It is found that graphical solutions based on the computerized Meta-Smith charts are fast, intuitive and accurate, indeed. In addition, basic procedures of using the Meta-Smith charts for solving CCITL and BCITL problems are similar to those of applying the standard Smith chart for solving problems associated with reciprocal uniform TLs. Thus, the Meta-Smith charts are useful in analysis and design of complicated CCITL and BCITL problems.

It should be pointed out that the Meta-Smith charts are applied to analyze and design *passive* circuits only in this book. However, they can be applied for problems associated with *active* circuits as well; e.g., active loaded TLs [10]–[11] and microwave transistor amplifiers in the CCITL

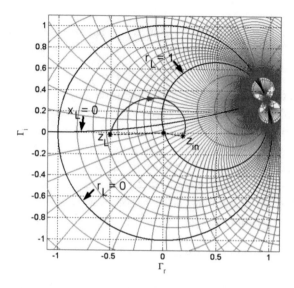

Figure 5.8: Impedance calculation on the Z Meta-Smith chart for BCITLs with NNCRs for $Z_0^+ = 149.3 - j7.9\ \Omega$, $Z_0^- = 146.9 - j39.6\ \Omega$ and $\gamma = 29.5 + j83.8$ 1/m.

system [12]–[14]. Further applications of the Meta-Smith charts will be illustrated in the future. Finally, it is hoped that the Meta-Smith charts will be used extensively by students, professors and researchers in the RF community in the near future to develop a physical understanding of many sophisticated phenomena of complex TLs and related problems.

BIBLIOGRAPHY

[1] D. M. Pozar, *Microwave Engineering*, 3rd ed. New Jersey: Wiley, 2005. 67, 68, 71

[2] A. Darawankuland D. Torrungrueng, "A single-stub tuning for nonreciprocal uniformtrans-mission lines," in *Proceedings of the First ECTI International Conference*, May 2004, vol. 1, pp. 83–86. 68, 70

[3] D. Torrungrueng and C. Thimaporn, "Applications of the ZY T-chart for nonreciprocal stub tuners," *Microwave and Optical Technology Letters*, vol. 45, no. 3, pp. 259–262, May 2005. DOI: 10.1002/mop.20789 68

[4] F. Urbani, L. Vegni, and A. Toscano, "A generalized Smith chart for an exponential tapered nonuniform transmission line," *Microwave and Optical Technology Letters*, vol. 14, no. 1, pp. 36–39, Jan. 1997. DOI: 10.1002/(SICI)1098-2760(199701)14:1%3C36::AID-MOP11%3E3.0.CO;2-A 73

[5] D. Torrungrueng and C. Thimaporn, "Application of the T-chart for solving exponentially-tapered lossless nonuniform transmission line problems," *Microwave and Optical Technology Letters*, vol. 45, no. 5, pp. 402–406, Jun. 2005. DOI: 10.1002/mop.20836 73, 74

[6] D. Torrungrueng, C. Thimaporn, and N. Chamnandechakun, "An application of the T-Chart for solving problems associated with terminated finite lossless periodic structures," *Microwave and Optical Technology Letters*, vol. 47, no. 6, pp. 594–597, Dec. 2005. DOI: 10.1002/mop.21239 76

[7] S. Lamultree and D. Torrungrueng, "On the characteristics of conjugately characteristic-impedance transmission lines with active characteristic impedance," in *2006 Asia-Pacific Microwave Conference Proceedings*, Dec. 2006, vol. 1, pp. 225–228. DOI: 10.1109/APMC.2006.4429411 77

[8] D. Torrungrueng, P.Y. Chou, and M. Krairiksh, "A graphical tool for analysis and design of bi-characteristic-impedance transmission lines," *Microwave and Optical Technology Letters*, vol. 49, no. 10, pp. 2368–2372, Oct. 2007. DOI: 10.1002/mop.22801 78

[9] D. Torrungrueng, P.Y. Chou, and M. Krairiksh, "Erratum: A graphical tool for analysis and design of bi-characteristic-impedance transmission lines," *Microwave and Optical Technology Letters*, vol. 51, no. 4, pp. 1154, Apr. 2009. 78

[10] C. Lertsirimit and D. Torrungrueng, "Analysis of active loaded transmission line using an equivalent BCITL model," in *2007 Asia-Pacific Microwave Conference Proceedings*, Dec. 2007, vol. 4, pp. 2353–2356. DOI: 10.1109/APMC.2007.4554830 79

[11] C. Lertsirimit and D. Torrungrueng, "Graphical tool in analysis of active loaded transmission lines," in *Proceedings of the 2009 ECTI International Conference*, May 2009, vol. 2, pp. 826–829. DOI: 10.1109/ECTICON.2009.5137173 79

[12] R. Silapunt and D. Torrungrueng, "A novel analysis of two-port networks in the system of conjugately characteristic-impedance transmission lines (CCITLs)," in *2005 Asia-Pacific Microwave Conference Proceedings*, Dec. 2005, vol. 3, pp. 1662–1665. DOI: 10.1109/APMC.2005.1606613 80

[13] R. Silapunt and D. Torrungrueng, "The VSWR considerations in potentially unstable microwave amplifiers design in the CCITL system," in *2007 Asia-Pacific Microwave Conference Proceedings*, Dec. 2007, vol.2, pp. 811–814. DOI: 10.1109/APMC.2007.4554840

[14] R. Silapunt and D. Torrungrueng, "An alternative approach in deriving associated circle equations for microwave amplifiers in the CCITL system," in *Proceedings of the 2008 ECTI International Conference*, May 2008, vol. 1, pp. 233–236. DOI: 10.1109/ECTICON.2008.4600415 80

5.8 PROBLEMS

5.1. Show that the input impedance of nonreciprocal lossless uniform OC or SC stubs is *purely imaginary*.

5.2. Design nonreciprocal lossless single-stub shunt tuners implemented using an OC stub as shown in Figure 5.1 (except that the stub is connected in shunt) when Z_L = 50 Ω, $|Z_0^\pm|$= 50 Ω, and $\phi = 45°$.

5.3. Design nonreciprocal lossless double-stub shunt tuners implemented using OC nonreciprocal stubs as shown in Figure 5.3 when Y_L = 10 + j50 S, $|Y_0^\pm|$= 50 S, and $\phi = 30°$, and $d = 3\,\tilde{\lambda}/8$. Compute the normalized conductance g_{max} defined in (5.6) as well. Interpret your result in terms of a forbidden region.

5.4. Verify that the equation of the conductance circle governing the forbidden region is given as in (5.5).

5.5. Using the Meta-Smith chart for CCITLs with NNCRs, compute the input impedance of the terminated ETLNUTLs with C_0= 30 pF/m, q = 2 Np/m, f = 5 GHz, L_0 = 1.688 nH/m, $\beta\ell = 45°$ and Z_L =7.5 Ω (defined in Section 5.3.2). Compare this graphical solution with the analytical one.

5.6. Verify the total *ABCD* matrix in (5.11).

5.7. In Section 5.4, the Meta-Smith chart for CCITLs with NNCRs is employed to determine the input impedance $Z_{in,M}$ in (5.12) of the terminated finite reciprocal lossless periodic TL structure of *three* unit cells in Figure 2.3, where each unit cell is given as in Figure 2.2. Using the Smith chart, determine $Z_{in,M}$ for this case, and compare with the result in (5.12). In addition, if the number of unit cells is increased to *six*, determine $Z_{in,M}$ using the Meta-Smith chart for CCITLs with NNCRs and the Smith chart, and compare these results. Is the Smith chart convenient to use in computing $Z_{in,M}$ for these cases? Explain in detail.

5.8. Using the Z Meta-Smith chart for CCITLs with NCRs, compute the normalized input impedance of the reciprocal CCITL with NCRs of length $\lambda/8$ with $\phi = 150°$, terminated in a normalized load impedance of z_L = 0.8 + j1.2 as shown in Figure 2.4, where λ is the wavelength of waves propagating along the CCITL.

5.9. In Section 5.6, the Z Meta-Smith chart for BCITLs with NNCRs is employed to determine the input impedance Z_{in} in (5.15) of the terminated reciprocal lossy unit cell in Figure 3.4. Using the Smith chart, determine Z_{in} for this case, and compare with the result in (5.15).

APPENDIX A

The Transmission (*ABCD*) Matrix

The transmission (*ABCD*) matrix is a useful representation to characterize a two-port network, especially for two-port networks with cascade connection. Figure A.1 illustrates a two-port network represented by the transmission (*ABCD*) matrix. The *ABCD* matrix is defined in terms of the *total*

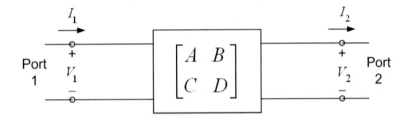

Figure A.1: A two-port network represented by the transmission (*ABCD*) matrix.

voltages (V_1 and V_2) and *total* currents (I_1 and I_2) as [1]

$$\begin{bmatrix} V_1 \\ I_1 \end{bmatrix} = \begin{bmatrix} A & B \\ C & D \end{bmatrix} \begin{bmatrix} V_2 \\ I_2 \end{bmatrix}. \tag{A.1}$$

Note that the *ABCD* matrix relates the voltage and current at port 1 of the network (V_1 and I_1) to those at port 2 (V_2 and I_2). From the defining relations of (A.1), each *ABCD* parameter can be determined as follows:

$$A = \left. \frac{V_1}{V_2} \right|_{I_2=0}, \tag{A.2}$$

$$B = \left. \frac{V_1}{I_2} \right|_{V_2=0}, \tag{A.3}$$

$$C = \left. \frac{I_1}{V_2} \right|_{I_2=0}, \tag{A.4}$$

$$D = \left. \frac{I_1}{I_2} \right|_{V_2=0}. \tag{A.5}$$

Note that A and C are *OC* parameters; i.e., they are determined by open-circuiting at port 2 (I_2 = 0). However, B and D are *SC* parameters; i.e., they are determined by short-circuiting at port 2 (V_2 = 0). It should be pointed out that A and D are dimensionless, while the units of B and C are in Ω and S, respectively. The *ABCD* parameters of some useful two-port circuits can be found in [1].

Figure A.2 shows a cascade connection of two two-port networks. The *ABCD* parameters of the first two-port network are A_1, B_1, C_1, and D_1, while those of the second two-port network are A_2, B_2, C_2, and D_2. Using (A.1), the relationships of the voltage and current at each port in

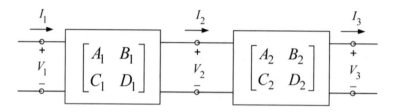

Figure A.2: A cascade connection of two two-port networks.

Figure A.2 can be written as

$$\begin{bmatrix} V_1 \\ I_1 \end{bmatrix} = \begin{bmatrix} A_1 & B_1 \\ C_1 & D_1 \end{bmatrix} \begin{bmatrix} V_2 \\ I_2 \end{bmatrix}, \tag{A.6}$$

$$\begin{bmatrix} V_2 \\ I_2 \end{bmatrix} = \begin{bmatrix} A_2 & B_2 \\ C_2 & D_2 \end{bmatrix} \begin{bmatrix} V_3 \\ I_3 \end{bmatrix}. \tag{A.7}$$

Substituting (A.7) into (A.6) yields

$$\begin{bmatrix} V_1 \\ I_1 \end{bmatrix} = \begin{bmatrix} A_1 & B_1 \\ C_1 & D_1 \end{bmatrix} \begin{bmatrix} A_2 & B_2 \\ C_2 & D_2 \end{bmatrix} \begin{bmatrix} V_3 \\ I_3 \end{bmatrix}. \tag{A.8}$$

This shows that the *ABCD* matrix of the cascade connection of two two-port networks is equal to the multiplication of the *ABCD* matrices of each two-port network. This result can be generalized to three or more two-port networks in a straightforward manner. In summary, the *ABCD* matrix of the cascade connection of two or more two-port networks is equal to the multiplication of the *ABCD* matrices of each two-port network, where the matrix-multiplication order must be identical to the order in which the networks are arranged. Thus, the *ABCD* matrix of more complicated networks consisting of cascades of simpler two-port networks can be readily found using the above property of the *ABCD* matrix.

BIBLIOGRAPHY

[1] D. M. Pozar, *Microwave Engineering*, 3rd ed. New Jersey: Wiley, 2005. 83, 84

APPENDIX B

The Required Condition for Nonreciprocal CCITLs

In this appendix, it will be shown that $\beta^+ \neq \beta^-$ is the required condition for *nonreciprocal* CC-ITLs (e.g., nonreciprocal lossless uniform TLs), where β^+ and β^- are the propagation constants of waves propagating in *forward* and *reverse* directions along nonreciprocal CCITLs, respectively. Using (2.1), (2.2) and the definitions of *ABCD* parameters given in Appendix A (see (A.2) to (A.5)), the transmission (*ABCD*) matrix of a nonreciprocal CCITLs of length ℓ as shown in Figure 2.1 can be determined as

$$\begin{bmatrix} A & B \\ C & D \end{bmatrix} = \frac{1}{Z_0^+ + Z_0^-} \begin{bmatrix} Z_0^+ e^{j\beta^+\ell} + Z_0^- e^{-j\beta^-\ell} & Z_0^+ Z_0^- \left(e^{j\beta^+\ell} - e^{-j\beta^-\ell} \right) \\ e^{j\beta^+\ell} - e^{-j\beta^-\ell} & Z_0^- e^{j\beta^+\ell} + Z_0^+ e^{-j\beta^-\ell} \end{bmatrix}. \tag{B.1}$$

For *nonreciprocal* CCITLs, it is required that the determinant of the *ABCD* matrix in (B.1) must *not* be equal to unity [1]. After some straightforward manipulations, it is found that

$$\det \begin{bmatrix} A & B \\ C & D \end{bmatrix} = e^{j(\beta^+ - \beta^-)\ell}. \tag{B.2}$$

From (B.2), the determinant of the *ABCD* matrix is not equal to unity if and only if $\beta^+ \neq \beta^-$ for an *arbitrary* length ℓ of nonreciprocal CCITLs. Thus, $\beta^+ \neq \beta^-$ is the *required condition* for *nonreciprocal* CCITLs indeed.

BIBLIOGRAPHY

[1] D. M. Pozar, *Microwave Engineering*, 3rd ed. New Jersey: Wiley, 2005. 85

APPENDIX C

Derivation of $Z_0^+ = (Z_0^-)^*$ for CCITLs

In Section 2.2, the traveling wave equations for the phasor voltage $V(z)$ and the phasor current $I(z)$ on CCITLs (e.g., nonreciprocal lossless uniform TLs) are given as in (2.1) and (2.2), respectively. Using the fact that CCITLs are *lossless*, the time-average power flow $P_{av}(z)$ along CCITLs at an arbitrary point z must be *constant*. By definition, $P_{av}(z)$ can be computed as

$$P_{av}(z) = \frac{1}{2}\text{Re}\left\{V(z)I^*(z)\right\}. \tag{C.1}$$

Substituting (2.1) and (2.2) into (C.1) yields

$$P_{av}(z) = \frac{1}{2}\text{Re}\left\{\frac{\left|V_0^+\right|^2}{\left(Z_0^+\right)^*} + \frac{V_0^-\left(V_0^+\right)^*}{\left(Z_0^+\right)^*}e^{j(\beta^+ + \beta^-)z} - \frac{V_0^+\left(V_0^-\right)^*}{\left(Z_0^-\right)^*}e^{-j(\beta^+ + \beta^-)z} - \frac{\left|V_0^-\right|^2}{\left(Z_0^-\right)^*}\right\}. \tag{C.2}$$

Using the fact that $P_{av}(z)$ is constant, the rate of change of $P_{av}(z)$ with respect to z must be equal to zero; i.e.,

$$\frac{d}{dz}\left(P_{av}(z)\right) = 0. \tag{C.3}$$

Substituting (C.2) into (C.3) yields

$$\text{Re}\left\{\frac{jV_0^-\left(V_0^+\right)^*}{\left(Z_0^+\right)^*}\left(\beta^+ + \beta^-\right)e^{j(\beta^+ + \beta^-)z} + \frac{jV_0^+\left(V_0^-\right)^*}{\left(Z_0^-\right)^*}\left(\beta^+ + \beta^-\right)e^{-j(\beta^+ + \beta^-)z}\right\} = 0. \tag{C.4}$$

Using the fact that β^+ and β^- are *real*, (C.4) can be rewritten compactly as

$$\text{Re}\left\{jT\right\} = 0, \tag{C.5}$$

where

$$T \equiv \frac{V_0^-\left(V_0^+\right)^*}{\left(Z_0^+\right)^*}e^{j(\beta^+ + \beta^-)z} + \frac{V_0^+\left(V_0^-\right)^*}{\left(Z_0^-\right)^*}e^{-j(\beta^+ + \beta^-)z}. \tag{C.6}$$

To satisfy (C.5), it is required that the term T must be *real*. Note that (C.6) can be rewritten as

$$T = \left(\frac{V_0^+\left(V_0^-\right)^*}{Z_0^+}e^{-j(\beta^+ + \beta^-)z}\right)^* + \frac{V_0^+\left(V_0^-\right)^*}{\left(Z_0^-\right)^*}e^{-j(\beta^+ + \beta^-)z}. \tag{C.7}$$

To make the term T *real*, it is obvious from (C.7) that

$$Z_0^+ = (Z_0^-)^*. \tag{C.8}$$

Thus, CCITLs possess $Z_0^+ = (Z_0^-)^*$ indeed.

APPENDIX D

Derivation of the Magnitude of the Voltage Reflection Coefficient at the Load for Terminated CCITLs

This appendix shows the derivation of the magnitude of the voltage reflection coefficient at the load, $|\Gamma|$, for terminated CCITLs, where Γ is defined as in (2.7). For convenience in derivation, Γ is rewritten in terms of normalized impedances as

$$\Gamma = (z_0^-)^2 \frac{(z_L - z_0^+)}{(z_L + z_0^-)}, \tag{D.1}$$

where z_L is the normalized load impedance and $z_0^\pm$ are the normalized characteristic impedances (see (4.4), (4.5), and (4.7)). Using the fact that $|z_0^-| = 1$ and defining $z_L \equiv r_L + jx_L$, $|\Gamma|$ can be expressed as

$$|\Gamma| = \left[\frac{(r_L - \cos\phi)^2 + (x_L + \sin\phi)^2}{(r_L + \cos\phi)^2 + (x_L + \sin\phi)^2} \right]^{1/2}, \tag{D.2}$$

where $r_L \geq 0$ for passive loads and ϕ is the argument of Z_0^-. In (D.2), when r_L is equal to zero, $|\Gamma|$ is always equal to one. In addition, it is obvious that the denominator will be greater than or equal to the numerator for the NNCR case since $\cos\phi \geq 0$. Therefore, $|\Gamma|$ is always less than or equal to unity for this case. On the other hand, $|\Gamma|$ is always greater than or equal to unity for the NCR case due to the fact that the numerator is always greater than or equal to the denominator when $\cos\phi < 0$.

Since $|\Gamma|$ can be greater than or equal to unity for the NCR case, one might wonder if this results in the violation of the conservation of power, especially when $|\Gamma|$ approaches infinity (when $z_L = -z_0^-$). For the NCR case, $|\Gamma|$ can approach infinity with *passive* loads ($\text{Re}\{z_L\} \geq 0$) due to the fact that $(\text{Re}\{z_0^-\} < 0)$. Appendix E shows that the associated power is still conserved for this case with passive load terminations.

APPENDIX E

Power Consideration for Terminated Finite Reciprocal Lossless Periodic Transmission Line Structures

Figure E.1 illustrates a finite reciprocal *lossless* periodic TL structure of M unit cells connected to a *passive* load impedance Z_L and a generator, where E_S and Z_S are the source voltage and the source impedance, respectively. The total time-average power P_m distributed along the mth terminal of the

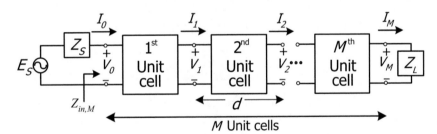

Figure E.1: A finite reciprocal lossless periodic TL structure of M unit cells connected to a passive load impedance Z_L and a generator.

periodic TL structure can be computed as

$$P_m = \frac{1}{2}\text{Re}\left\{V_m I_m^*\right\}, \tag{E.1}$$

where the phasor voltage V_m and the phasor current I_m in the passsbands at the terminal of the mth unit cell (where $m = 1, 2, \ldots, M$) can be written as (see (2.16), (2.17), and (2.27))

$$V_m = V_0^+ e^{-jm\beta d} + V_0^- e^{jm\beta d}, \tag{E.2}$$

$$I_m = \frac{V_0^+}{Z_0^+}e^{-jm\beta d} - \frac{V_0^-}{Z_0^-}e^{jm\beta d}, \tag{E.3}$$

where $Z_0^{\pm}$ are the effective characteristic impedances and β is the effective propagation constant of the periodic TL structure. Substituting (E.2) and (E.3) into (E.1), P_m in passbands can be expressed compactly as follows [1]:

$$P_m = \frac{|V_0|^2}{2\,|Z_0|} \frac{\left(1 - |\Gamma|^2\right)}{|1 + \Gamma e^{-j2M\beta d}|^2} \cos\phi,$$ (E.4)

where the voltage reflection coefficient at the load Γ in (E.4) is defined as

$$\Gamma \equiv \frac{V_0^- e^{jM\beta d}}{V_0^+ e^{-jM\beta d}} \equiv \Gamma_0 e^{j2M\beta d}.$$ (E.5)

Note that Γ_0 is the voltage reflection coefficient at the input, where Γ_0 is defined as

$$\Gamma_0 \equiv V_0^- / V_0^+,$$ (E.6)

and the input voltage V_0 can be written in terms of Γ_0 as

$$V_0 = V_0^+ \left(1 + \Gamma_0\right).$$ (E.7)

Alternatively, V_0 can be determined from the circuit in Figure E.1 as

$$V_0 = \frac{Z_{in,M}}{Z_{in,M} + Z_s} E_s,$$ (E.8)

where $Z_{in,M}$ is the input impedance of the terminated periodic TL structure. It can be seen from (E.4) that, for both NNCR ($|\Gamma| \leq 1$) and NCR cases ($|\Gamma| \geq 1$), P_m is still *nonnegative* since $\cos\phi \geq 0$ for the NNCR case and $\cos\phi < 0$ for the NCR case.

From (E.4) and (E.8), if $|\Gamma|$ approaches infinity (when $z_L = -z_0^-$ as discussed in Appendix D) , P_m becomes [1]

$$P_m = -\frac{1}{2} \frac{|Z_{in,M}|^2}{|Z_0|} \cos\phi \left|\frac{E_s}{Z_{in,M} + Z_s}\right|^2,$$ (E.9)

which is *nonnegative* as well since $\cos\phi < 0$ for the NCR case. Thus, the power is still *conserved* for the NCR case with passive load terminations even when $|\Gamma|$ approaches infinity. For a fixed passive load impedance and a fixed generator, P_m is *constant* along each terminal of the periodic TL structure operating in passbands since the structure of interest is *lossless*.

BIBLIOGRAPHY

[1] S. Lamultree and D. Torrungrueng, "On the characteristics of conjugately characteristic-impedance transmission lines with active characteristic impedance," in *2006 Asia-Pacific Microwave Conference Proceedings*, vol. 1, pp. 225–228, Dec. 2006. 92

Author's Biography

DANAI TORRUNGRUENG

Danai Torrungrueng received his B.Eng. degree in electrical engineering from Chulalongkorn University, Bangkok, Thailand, in 1993. He obtained his M.S. and Ph.D. degrees in electrical engineering from The Ohio State University in 1996 and 2000, respectively. From 1995 to 2000, he was a Graduate Research Assistant (GRA) in the Department of Electrical Engineering, ElectroScience Laboratory of The Ohio State University. Prior to joining Asian University, he worked as a senior engineer in the USA, involved in research and development of the urban propagation modeling project. At present, he is an associate professor in the Electrical and Electronic Engineering Department in the Faculty of Engineering and Technology at Asian University, Thailand.

In 2000, he won an award in the National URSI Student Paper competition at the 2000 National Radio Science Meeting in Boulder, Colorado. During 2004 to 2009, he invented and developed Meta-Smith charts (*visit* http://www.meta-smithcharts.org) for solving problems associated with conjugately characteristic-impedance transmission lines (CCITLs) and bi-characteristic-impedance transmission lines (BCITLs) graphically. His research interests are in the areas of fast computational electromagnetics, antenna design, RFID, rough surface scattering, propagation modeling and electromagnetic wave theory. He is currently a senior member of the IEEE and a member of the Electrical Engineering/Electronics, Computer, Telecommunications and Information Technology Association of Thailand (ECTI).

Index